RATIONAL METRIC NOTATION

Other Books by the Same Author

Principles of Rhythm
Creative Harmony

RATIONAL METRIC NOTATION

The Mathematical Basis of
Meters, Symbols, and Note-Values

Paul Creston

An Exposition-University Book

EXPOSITION PRESS   HICKSVILLE, NEW YORK

CONTENTS

v

PREFACE

For over 400 years an equivocal and imprecise metric terminology has been preached, unquestioned and unchallenged by students or teachers. This noumenon has been for too long bemoand by me in all my lectures and composition classes. At last, the words of Confucius came to my mind: "Better to light one candle than to curse the darkness." The result of that advice is this book, which purports to light the candle of logic to dispel the darkness of obscurantism. The full title might be: *The Theory and Practice of a Modern Rational Metric Notation and Terminology*; it is a much fuller treatment of the sections on Meter and Metric Notation than in my previous book, *Principles of Rhythm*.

After an examination of over fifty definitions of Rhythm, Curt Sachs (*Rhythm and Tempo*, 1953) remarked: "The confusion is terrifying indeed." Were he alive today to examine some of the latest definitions, his remark would be a feeble complaint. And when we turn our attention specifically to Meter, and to past and present dicta, the remark would be the understatement of the century. Some avant-garde composers have thickend the turbidity and heightend the confusion thru their ingraind misconceptions and concocted paralogisms that place the performer in an incongruous situation. For while scientists have attempted for many decades to invent machines that can think and act like human beings, avant-garde rhythmists are attempting to develop musicians—human beings—who can think and act like machines. If musical rhythms must be calculated by slide rule or computer, the human, intuitive factor in art is completely vitiated.

The two major causes of the confusion are (1) the ambiguous terminology of note-values, and (2) the misconception regarding the function of the barline. The solution to the problem is incredibly simple, and the conversion to a rational metric system can be effected by an equally simple device. Also, it must be understood that no *new* symbols of notation have been proposed, for no *invented* system can succeed as well as one that has evolvd, that has develpt from the known to the unknown.

In 1912, the three linguistic societies of America officially approved the Revised Scientific Spelling of the English language.* To this day, more than a half-century later, this alphabet has not been adopted by any writer or editor of note, or by the educational system. The prospect, consequently, of general or early acceptance of a Rational Metric Notation seems discouragingly dim, but does not minify the urgency of such a proposal or deter the Author in presenting it.

The reader may well ask for whom the book is intended. At the risk of seeming vainglorious, I must declare that it is intended for everyone interested in the theory or practice of music, from the theory student to the musicologist, and from the harmonica player to the conductor.

*   *   *

*Revised Spelling has been adopted only in past tense words in which the 'e' of 'ed' is silent. When the 'e' modifies the sound of the preceding consonant, it is retained, e.g. in 'unchallenged.' Also, in such words as 'paradigm,' the silent 'g' is omitted.

RATIONAL METRIC NOTATION

I
PRINCIPLES OF RHYTHM— SUMMARY

One must distinguish that which is traditional because it is right from that which is right only because it is traditional.

Meter and Rhythm are not synonymous. Meter is only one of the four elements of Rhythm, the other three being: Pace (Tempo), Accent and Pattern. To understand what is properly in the domain of Meter we must have a clear conception of all four elements and the function of each. This chapter is an epitome of the principles of Rhythm as more fully exposed in the Author's book on that subject.

RHYTHM, in music, is the organization of duration in orderd movement. An indefinitely sustaind tone manifests simple duration—not organized; the duration must be organized into equal or unequal segments in order to manifest Rhythm. The glissando of the fire siren exemplifies simple movement—not orderd. Hence the importance of the words 'organization' and 'orderd' in the definition of musical rhythm.

In a single measure, the pulses—commonly calld 'beats'—are the units of duration; in a phrase, the measures themselves are considerd the units; and in an entire composition, the formal sections are the units.

ELEMENTS OF RHYTHM

The four elements of Rhythm—Meter, Pace, Accent and Pattern—are the primary organization of duration (the irreducible min-

1

imum in the analysis of any rhythm) and a change in any one of the elements results in a change of rhythm. As an illustration, let us alter each element in the rhythm of Bach's Passacaglia for Organ, which is in 3/4 meter, slow pace, agogically accented on the first pulse, and in an iambic (short-long) pattern.

EXAMPLE 1.

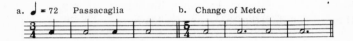

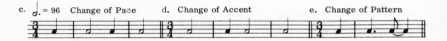

By changing the 3/4 to 5/4 (Example 1b) there results a different organization of duration and a different order of movement. The change to a fast pace (Example 1c) converts it to a Scherzo or Tarantella. Shifting the agogic accent to the second pulse transforms the Passacaglia into a Sarabande (Example 1d); and the pattern of Example 1e creates what is termd the 'hemiola' rhythm. In short, the four elements are the irreducible minimum in the analysis and description of *any* rhythm.

The logic of the very first sentence should be evident: "Meter and Rhythm are not synonymous." This chapter's summary of rhythmic principles will further clarify that concept.

METER

The ensuing chapters will be devoted to the full exposition of Meter as practiced in the music of Western civilization from 1600 to the present day. In this chapter it will be merely sketcht sufficiently to show its relationship to, and function in, rhythmic practice.

Meter is the grouping of pulses or units in a measure or a frame of two or more measures. Another definition could be: Meter is the measurement of duration by pulses or units. The word 'meter' itself is derived from the Greek 'metron,' the Latin 'metrum' and the French 'mètre'; in music, it has only *one* meaning—'measure.'

The Elements of Meter are: pulses, beats and units. PULSES —or 'metrical pulses'—are commonly termd 'beats': 2/4 has 2 pulses, 3/4, 3 pulses, etc. The term 'beat' is reservd for the actual (rhythmic) beat which may or may not coincide with the metrical pulse. In the following example, the metrical pulses and the rhythmic beats are the same or synchronous.

EXAMPLE 2

In Example 3, however, there are 3 pulses of quarter-note-value each but 2 beats of 3/8-note-value each; the 3/8 note is commonly but imprecisely termd a 'dotted-quarter' note; this will be fully explained under Note-Values.

EXAMPLE 3

$$\frac{3}{4} \quad \text{BEATS} \quad \text{PULSES}$$

UNITS are the subdivision of a pulse or beat into smaller fractions of duration. The units in Example 3 would be 8th notes, so the rhythm can be diagramd as follows:

EXAMPLE 4

There are numerous divisions of a pulse or beat found in music of Western civilization. They will be fully considerd later.

PACE, commonly termd 'Tempo,' is the rate at which the pulses or units of a meter occur, noted by the words *Lento, Moderato, Allegro,* etc., or by a metronomic indication: ♩ =48, ♩ =96, ♪=60, etc. In Italian, the word 'tempo' has *four* different musical meanings: pulse, meter, movement (of a Sonata, Symphony, etc.) and pace. What musicians call 'tempo' is usually termd, in Italian, 'andamento.' Hence the insistence on the more precise term 'pace.'

In the classification of meters, Pace is correlated to Meter. For example, in 3/4 meter at a pace of ♩ =96 (Example 5a) the 1/4 notes are the pulses and the 1/8 notes, the primary units; the meter would be classified as 'Binary,' that is, 2 primary units to a pulse. At a pace of ♩. =96 (Example 5b), the entire measure (3/4 note-value) is the pulse and the 1/4 notes become the primary units, transforming the 3/4 Binary meter to Ternary (3 primary units to a pulse).

EXAMPLE 5

Pace is governd by 2 factors: the length of the meter and the units employd in the pattern. A long meter (4/4 or 5/4) gives the impression of a slower pace than a short meter (2/4), even tho the actual rate of pulses may be the same. This is due to the more widely spaced primary pulses, as the following experiment with simple counting or beating will demonstrate.

EXAMPLE 6

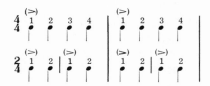

The second factor, the units employd in a pattern, also alters the pace. A large unit—whole-note or half-note—gives the impression of slower pace than a small unit—1/8-note or 1/16-note—the pulse-rate being constant.

EXAMPLE 7

This second factor reveals also the correlation of *Pace* to *Pattern,* as well as its correlation to *Meter.*

ACCENT is that element of rhythm which emphasizes a pulse or beat. There are at least 8 types of rhythmic accent: (1) Dynamic, (2) Agogic, (3) Metric, (4) Harmonic, (5) Weight, (6) Pitch, (7) Pattern, and (8) Embellisht. A 9th type that might be labeld Tone-Color or Orchestral is not included because it rightly belongs in orchestration and also because it is actually a form of Weight accent. The Orchestral accent is effected by a stroke on the triangle, timpanum, bass drum or any percussion instrument. Another type of accent which is not included is the Expressive accent because it is nonrhythmic, that is, it does not affect or alter the rhythm. The Expressive accent emphasizes a single tone in a run or florid passage.

Most of the accent types are usually present in combination with another or others, depending on the degree of accentuation or number of simultaneous accents required by the musical expression. The 8 types of accent are now explaind.

The DYNAMIC accent, the most common type, is a qualitative accent and emphasizes a pulse or beat by means of tonal intensity. It is noted in varying degrees by certain symbols: the macron (-) for the slightest degree, >, ∧, *sf*, and *sff*—the greatest degree. Of all the types, the dynamic accent is the only one that is supplied by the performer instead of being in the rhythmic structure itself.

EXAMPLE 8

The AGOGIC accent is a quantitative accent and emphasizes a pulse or beat durationally, that is, by means of a tone longer than the preceding one, and/or the following tone. It is employd most frequently in lyric melodies and in organ music. (Arrows indicate the accented tone.)

EXAMPLE 9

The METRIC accent is implied in the uniform grouping of pulses with one harmony to each measure.

EXAMPLE 10

The HARMONIC accent emphasizes a pulse or beat by means of a dissonance on that pulse or beat. In Example 11 the Harmonic accent is combined with the Agogic since the accented tone is followed by a shorter tone—in the first measure—and both preceded and followed, in the second measure.

EXAMPLE 11

The WEIGHT accent expresses emphasis thru texture or volume—in terms of mass, not intensity or loudness.

EXAMPLE 12

The PITCH accent is present at the highest or lowest tone of a group. The most common type is the accompaniment to a waltz or polka, but it is also apparent in florid passages.

EXAMPLE 13

The PATTERN accent is evident in a repeated figure of characteristic contour. This type functions like the Agogic in that it emphasizes a pulse or beat without tonal intensity. In Example 14 it is combined with the Pitch accent.

EXAMPLE 14

The EMBELLISHT accent is self-explanatory, effected by any form of ornament—grace note, mordent, appoggiatura, trill, turn (traditional or any group), etc.

EXAMPLE 15

PATTERN is the subdivision of a pulse, beat or measure into smaller units—equal or unequal. For example, a quarter-note pulse or beat ♩ ⌡may have a 2-note pattern: ♫ , ♪♪ , ♫ , ♫ , etc.; or a 3-note pattern: ♫♪ , ♫♫ , ♫♫ , etc.; or a pattern of any number of notes. A measure of 3/4 may have a pattern of a 3/4 note ♩. : ♩ ♫♫ | ♫♩ ♫♫ | etc.

The element of Pattern in music disproves the analogy of musical meter to poetic meter, which so many writers have proposed. For while there are, in poetic meter, 2 or 3 syllables to a foot, there are innumerable patterns of units to a *pulse,* the equivalent of the poetic *foot.*

RHYTHMIC STRUCTURES

The organization of duration in orderd movement is ultimately accomplisht by five different plans termd Rhythmic Structures. These are: I. Regular Subdivision, II. Irregular Subdivision, III. Overlapping, IV. Regular Subdivision Overlapping, and V. Irregular Subdivision Overlapping.

FIRST STRUCTURE: Regular Subdivision

Regular Subdivision is the organization of a measure into equal beats, that is, beats of equal duration, different than the number of pulses. This principle of rhythm stems from Meter itself, incorporating a number of *equal* beats against a different number of equal *pulses*. The most common and elemental form of this structure is the 'hemiola': 3 pulses—2 beats, or 2 pulses—3 beats.

EXAMPLE 16

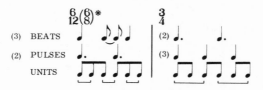

(*Metric signatures in parentheses are the traditional nomenclature. The revised notation (6/12) will be explaind later.)

Several compositions rhythmically based on the hemiola:

>   Chopin—Waltz in A-flat, Op. 42,   Scherzo, Op. 54
>   Brahms—Capriccio, Op. 76 No. 8,   Capriccio, Op. 76
>   No. 5
>   Bach—Fugue No. 10, WTC Book I
>   Schumann—Aufschwung, Op. 12
>   Massenet—Aragonaise.
>   Ravel—Valse No. 7

Most of the examples of the 5 Rhythmic Structures will be found in the Author's *Principles of Rhythm.*

Were Regular Subdivision possible only in 6/12 and 3/4, it would not constitute a principle. It is, in fact, applicable to all meters, and in several forms other than the hemiola. Following are two other forms of the First Structure, in 6/12 and in 3/4:

EXAMPLE 17

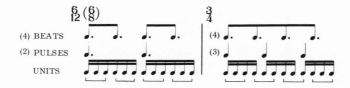

And the First Structure in other meters:

EXAMPLE 18

The phenomenon of Rhythmic Structures justifies the three elements of Meter: Pulses, Beats and Units.

Several sources of the First Structure:

2 pulses-4 beats	Debussy—Les collines d'Anacapri
	Chopin—Ballade, Op. 38
	Scriabin—Prelude, Op. 56 No. 1
3 pulses-4 beats	Chopin—Valse, Op. 64 No. 1 (measure 9—3rd section)
	Villa-Lobos—O loboshinho de vidro
4 pulses-3 beats	Creston—Invocation and Dance (measures 118-130)

4 pulses-6 beats   Brahms—Rhapsody, Op. 79 No. 2
Villa-Lobos—O gatinho de papelão
Villa-Lobos—O boisinho de chumbo
Creston—Prelude, Op. 38 No. 2

The Creston Prelude No. 2—from Six Preludes for Piano—is based entirely on the First Structure.

SECOND STRUCTURE—Irregular Subdivision

Irregular Subdivision is the organization of a measure into unequal beats, i.e. beats of differing duration. The principle is inherent in compound meters like 5/4 and 7/4, which are generally treated as, respectively, 3/4 + 2/4 and 4/4 + 3/4. The most common form of the Second Structure is the 3 + 3 + 2 pattern in 4/4, the basic rhythm of the Cuban rumba.

EXAMPLE 19

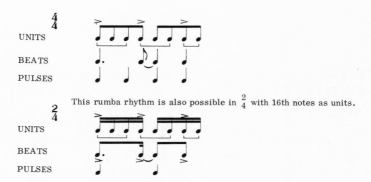

This rumba rhythm is also possible in $\frac{2}{4}$ with 16th notes as units.

The principle of Irregular Subdivision is applicable in all meters. Study the following partial tabulation.

EXAMPLE 20

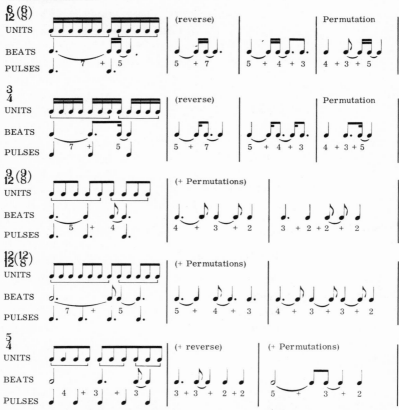

Permutations are possible with any series of beats, e.g. 3 + 3 + 2 can be rearranged as 3 + 2 + 3 or 2 + 3 + 3. Also, in the Second Structure, there can be the same number of pulses as beats, e.g. in 12/12 = 4 pulses and 4 irregular beats.

Several sources of the Second Structure:

3 + 5 and 3 + 3 + 2 in 4/4	Villa-Lobos—O ga-tin de papelão
3 + 3 + 2 and 3 + 2 + 3 in 8/8	Ravel—Trio-Piano, Violin and Cello (1st movement)

3 + 3 + 2 only   Bartok—Mikrokos-
                                         mos No. 153
                                         Bartok—String Quar-
                                         tet No. 4
                                         Creston—Symphony
                                         No. 4 (1st move-
                                         ment)
4 + 3 + 2 in (9/8)   Creston—Symphony
                                         No. 4 (2nd move-
                                         ment)
5 + 4 in (9/8) written 5/8 + 4/8   Scriabin—Prelude
                                         Op. 11 No. 16
4 + 2 + 3 in (9/8) written 4 + 2 + 3   Bartok—Mikrokos-
                                    ――― mos No. 148
                                     8

All the above examples are found in *Principles of Rhythm.*

THIRD STRUCURE—Overlapping

Overlapping is the extension of a phrase rhythm beyond the barline; that is, changing meters without changing the metric notation, Schuman's *Träumerei* is notated in 4/4 but the actual rhythmic phrasing (notated by dotted barlines) is: 5/4—4/4—2/4 —2/4—3/4.

EXAMPLE 21

The Third Structure was common practice in the 16th century, virtually abandond for 300 years, and re-adopted in the 20th century. In its adoption in the 20th century, however, the tyranny of the barline—the misconception that the first pulse of a measure is always strong—was irrationally maintaind. In the following typical specimen of 16th-century rhythm the entire motet is written in 4/4 despite the changing meters (noted by dotted barlines). A 20th-century composer would probably indicate the changes by **new** metric notations.

EXAMPLE 22

Ecce quomodo moritur — Jacobus Gallus (Jacob Händl)

In *Principles of Rhythm* will be found a number of examples in 20th-century music that justify the definition of 'Meter' as a grouping of pulses or units without accentual value, thereby making constant metric changes in notation entirely unnecessary. The examples are chosen from the music of Debussy, Piston, Bartok, Walton, Creston, et al.

FOURTH STRUCTURE—Regular Subdivision Overlapping

Regular Subdivision Overlapping is a combination of the First and Third Structures and is defined as: the organization of a *group of measures* into equal beats *overlapping* the bar. In other words, it involves two or more measures. As the hemiola is the most common form of the First Structure, so is it of the Fourth, this time in 2 measures of 3/4 (or 3/2 or 8/8).

EXAMPLE 23

Note: In this structure the *measures* become the pulses, and the *pulses* become the units.

Several examples of the Fourth Structure in various meters:

EXAMPLE 24

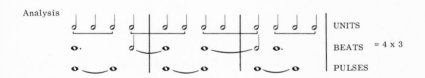

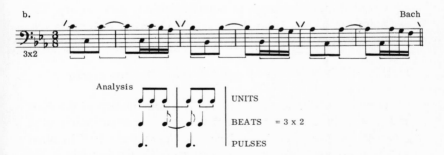

Note: Combination of Pitch and Pattern accents.

c. Scherzo from "A Midsummer Night's Dream" - Mendelssohn

3x4

Analysis   UNITS

BEATS   = 3 x 4

PULSES

Note: Combination of Embellisht and Agogic accents.

d.   2nd movement - Symphony No. 6 ("Pathetique") -
                                                    Tchaikovsky

5x2
Analysis   UNITS

BEATS   = 5 x 2

PULSES

e.
                    4th movement - Symphony No. 2 - Brahms

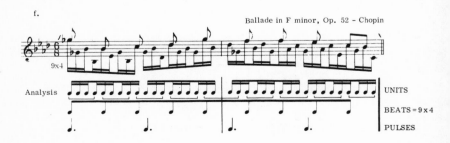

Analysis

4x3   UNITS

BEATS   = 4 x 3

PULSES

f.
                            Ballade in F minor, Op. 52 - Chopin

9x4

Analysis   UNITS

BEATS = 9 x 4

PULSES

The Chopin example (24f) is a master stroke of a rhythm genius. It is ingenious in the calculation and in the practicality of execution; the melody in 9th notes and, on the last beat, 18th notes—written as 8th notes—can be played only by a pianist, the fingers measuring the 36th notes, written as 16th-note triplets. In the chapter on note-values, ninths, eighteenths, and thirty-sixths notes will be fully explained. Many examples of the Fourth Structure will be found in *Principles of Rhythm*.

FIFTH STRUCTURE—Irregular Subdivision Overlapping

Irregular Subdivision Overlapping is the organization of a group of measures into *unequal* beats overlapping the bar. It is a combination of the Second and Third Structures, and is a rhythmic phenomenon particularly of the 20th century.

The difference between the Fifth Structure and the Third is the presence—in the Fifth Structure—of a repeated rhythmic pattern. If in 2 measures of 2/4 we find the 3 + 3 + 2 rhythm *not* repeated, it would be termd simple Overlapping (Third Structure). On the other hand, if that same rhythm is repeated, it would be termd Irregular Subdivision Overlapping (Fifth Structure).

Following are specimens of the Fifth Structure with an analysis and the source of each given:

EXAMPLE 25

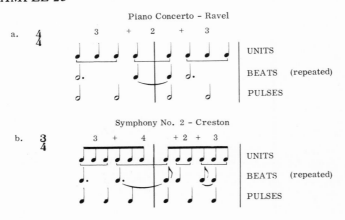

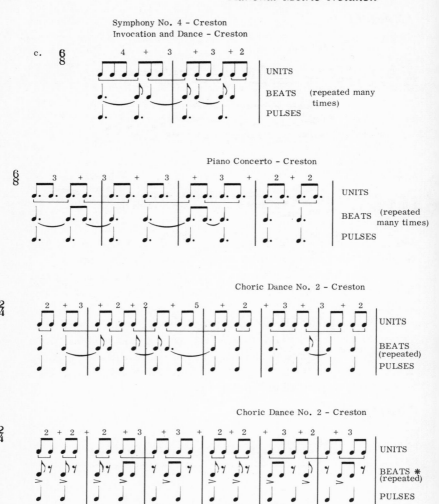

Symphony No. 4 - Creston
Invocation and Dance - Creston

Piano Concerto - Creston

Choric Dance No. 2 - Creston

Choric Dance No. 2 - Creston

* Pattern is included - the accented notes are the actual beats.

The Fifth Structure is sometimes notated as a Sequential meter, especially when the entire composition is based on that sequence. As Sequential meter it can also be found in pre-20th century music.

EXAMPLE 26

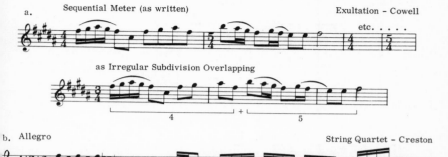

a. Sequential Meter (as written)   Exultation - Cowell

etc. . . . .

as Irregular Subdivision Overlapping

b. Allegro   String Quartet - Creston

Other specimens of the Fifth Structure will be found in Chapter Seven of *Principles of Rhythm*, as are the preceding examples.

To conclude this summary of rhythmic principles, mention must be made of Multimeters, Polymeters, Multirhythms and Polyrhythms.

A MULTIMETER is one in which the meter is changed continuously or continually within a section of a composition. Since Meter is a measurement of duration, the changes must be in the length of the meter, that is, from 2 pulses to 3 pulses, 3 to 2, etc., or from one number of units to another.

EXAMPLE 27

a. *Still It Frieth* — Morley

Yet my heart never di-eth, my heart never di-eth, my heart never di - eth.

b. *Fugue No. 4, Book II, Well-Tempered Clavier* — J. S. Bach

Multimeters is the term used in 20-century music for the principle of Overlapping (Third Structure), and many examples will be found in works by Bartok, Stravinsky, Prokofieff, et al.

A POLYMETER is one in which 2 or more meters are combined simultaneously.

EXAMPLE 28

Minuet from Don Giovanni — Mozart

The 3 meters (so notated) in the Mozart Minuet can, thru the
principle of Overlapping, be notated in 3/4, as the dotted-barlines
show.

A MULTIRHYTHM is one that contains successively different structures (Example 29a) or different configurations of the
same structure (Example 27b). Roman numerals designate the
rhythmic structure.

EXAMPLE 29

a. 3rd Movement of Symphony No. 4 — Creston

Property of G. Ricordi & Co., New York.

b. 1st Movement of Symphony No. 2 — Creston

Copyright 1954. Used by permission of G. Schirmer, Inc.

A POLYRHYTHM is one in which 2 or more structures, or
2 or more configurations of the same structure are combined simultaneously.

EXAMPLE 30

Walt Whitman, Op. 53 – Creston

II
ELEMENTS OF METER

Apathy toward semantics engenders imprecise terminology and shalloe thaut.

Meter has been defined as "the grouping of pulses or units in a measure or a frame of 2 or more measures"; also, as "the measurement of duration by pulses or units." Until the 19th century there were 2 systems of subdividing pulses: (1) into 2 primary units as in 2/4, 3/4, and 4/4; and (2) into 3 primary units as in traditional (6/8), (9/8) and (12/8). In the 20th century, a third system has been employd: measurement by units, in such meters as 7/16, 11/16 and 13/16, which are therefore named Unitary meters.

The elements of Meter are: Pulses, Beats and Units, as explaind in chapter I. Accent is *not* an element of Meter since it does not affect in any degree the *measurement* of duration. But Pace *does* alter duration as can be realized in a piece playd at different metronomic markings.

Curt Sachs (*Rhythm and Tempo*) wisely reminds us that "Meter has the backing of a tradition strong and good enough to justify its use for *rhythm by length,* and *nothing but rhythm by length.*" How far theorists and composers have strayd from this clear and sound concept can be noted from the following specimen definitions with this Author's comments.

1. "[Meter is] the part of rhythmical structure concernd with the division of a musical composition into measures by means of *regularly recurring accents* with each measure consisting of a uniform number of beats or time units, the *first of which has the strongest accent.*" (*Webster's Third New International Dictionary*). COMMENT: Accent is an element of rhythm not meter; meter is not governd in any way by accent. The Waltz, the Sarabande and the Mazurka are all in triple meter—3/4 or 3/2; but

the Waltz is accented on the first pulse, the Sarabande, on the second, and the Mazurka, on the third. Musical examples are numberless in which the accent falls on the 2nd, 3rd or 4th pulse of a 4/4 meter. Furthermore, all music based on the Third Structure—Overlapping—in which the note on the first pulse of a measure is tied to the preceding measure, or in which there is a rest on the first pulse, repudiates any relationship of Accent to Meter.

2. "Meter. The basic scheme of *note-values* and *accents* which remains *unaltered* throughout a composition or a section thereof and which serves as a skeleton for the rhythm" (*Harvard Dictionary of Music, 1964*). COMMENT: Only dull and sterile hymn tunes would be built on an *unaltered* basic scheme of note-values, such as 4 quarter notes in every measure of 4/4. If the author of that definition means by 'note-values,' 'pulses' or 'units,' he is forgiven the first sin—'basic scheme of note-values'—but still remains guilty of the second sin—'accents.'

3. "Meter is . . . traditionally speaking . . . a recurring *pattern of stress,* and an established arrangement of *strong and weak pulsations*. These pulsations are also known as *beats*" (*Music Notation,* Gardner Read). COMMENT: The italicized words confuse accent as an element of Meter instead of Rhythm.

4. "Meter is the measurement of the number of pulses between more or less *regularly recurring accents*" (*The Rhythmic Structure of Music,* Cooper and Meyer). COMMENT: Half-true and half-false; 'measurement,' true, 'accents,' false.

5. "Because the impression of rhythm depends not only upon the existence of *accented and nonaccented beats* but also upon the *grouping of those beats,* meter can in a sense exist alone without any impression of rhythm" (*Emotion and Meaning in Music,* Leonard B. Meyer). COMMENT: Meter *can* exist alone without any impression of rhythm—*in a tabulation*—but not in a composition. In a composition, the metric signature must be accompanied

by an indication of *Pace*—slow, moderate, fast, etc. If rhythm is the organization of duration in orderd movement, then, in a sense, meter certainly gives the impression of rhythm because the *duration* of a meter is *organized* and the *movement orderd* by the pace and pulses. Furthermore, as stated earlier, the First Rhythmic Structure—Regular Subdivision—stems from Meter as the organization of a measure into equal *beats* different than the number of *pulses*. The statement, however, relating accents and nonaccents to 'rhythm,' not 'meter,' is true.

Since we are concernd with fallacious definitions, we should consider also the common misconceptions regarding pulses, beats, units, bar, barline, and other metric terms.

PULSES AND BEATS

The first and principal definition of 'beat' given by Webster is "a stroke or blow, *especially one producing sound.*" In music, all the pulses of a measure may or may not be sounded, but beats *must* be sounded. An examination of Example 22 (q.v.) will show that, of the 44 measures in 4/4, only *one* (the third from the end) sounds *all* the pulses of the measure. By means of rests or of ties overlapping the bar, certain pulses are silent. It is true that in the simplest music, hymn tunes and folksongs, as noted earlier, pulses, beats and even units may be synchronous, and no distinction can be made. But in a great deal of art music the three elements of Meter are clearly present. Therefore, the term *beat* should be reservd for the actual *sounded* note or chord in all extrametrical rhythms, that is, rhythms based on any of the rhythmic structures. (The term 'extrametrical' means: not indicated in the metric signature.) Even in strictly metrical rhythms such as in the Turkish March (Example 31), both pulses and beats are present in the third measure. The Turkish March is in 2/4; the second pulse of the first, second and fourth measures is silent, while the first pulse of each measure is sounded and is consequently also a beat. Traditional terminology has never taken note of this distinction, and the latest writings still insist on naming the division of a measure 'beats' or 'counts,' instead of 'pulses.' Note how in Mozart's Rondo alla Turca and in Beethoven's Turkish March, the beats are expressed by means of Pitch accents.

EXAMPLE 31

UNITS are the subdivision of a pulse or beat into smaller fractions of duration than the pulse or beat. The fractions of a pulse are termd 'metrical' units and those of a beat, 'rhythmic' units. Metrical units are further classified—in order of greater fractionalization—as primary, secondary, tertiary and quartan. Units as an element of Meter are also necessary in order to distinguish between binary (simple) and ternary (compound) subdivisions of the pulse. Binary meters are those in which the primary units—the first subdivison of the pulse—are 2 to a pulse. Ternary meters contain 3 primary units to a pulse. The traditional terms 'simple' for binary and 'compound' for ternary will be shown to be imprecise. The following illustration clarifies the terminology of metrical units. Metric signatures in parentheses, in this book, are always the traditional nomenclature.

EXAMPLE 32

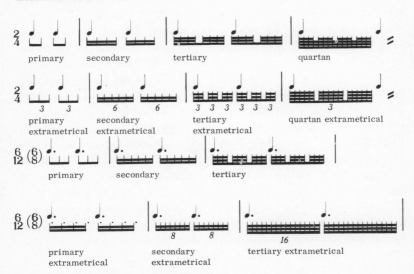

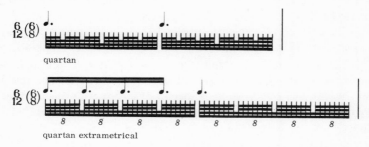

quartan

quartan extrametrical

Unit groups of 5, 7, 11, 13, etc., which are irregular in both Binary and Ternary meters, must be named simply by their fractional value. For example, a quintuplet equal to a quarter note (5 units to a pulse) would be termd 5/20 units. The calculation of irregular note-values is fully explained in the next chapter.

TYPES OF METER

BINARY AND TERNARY

The traditional terms 'simple' for binary and 'compound' for ternary are imprecise, equivocal, illogical and deceptive. Traditionally, 5/4 should be calld a 'simple' meter and (6/8), 'compound.' But there is nothing *simple* about 5/4,* which is generally treated as 3/4 + 2/4 or 2/4 + 3/4, in other words, as *compound;* and there is nothing *compound* in (6/8), which consists of 2 pulses subdivided into 3 primary units to a pulse. In medieval theory, meters subdivisible into 2 primary units to the pulse were classified as 'prolation imperfect,' and those meters containing 3 primary units, 'prolation perfect.' To later theorists those terms seemd equivocal, and 'simple' was substituted for 'prolation imperfect,' and 'compound' for 'prolation perfect.' But what makes the number '2' imperfect or simple, and the number '3' perfect or compound? (To digress for a moment, the word 'perfect' should not be used even in harmonic terminology. If the intervals of the 4th and 5th are perfect, then the tritone [named augmented

---

*In American music theory, 5/4 is termed 'mixed'; in Italian theory it is called 'composto' (compound).

4th or diminisht 5th] should be calld 'imperfect 4th' or 'imperfect 5th.' Some medieval theorists justified the use of 'perfect' for the number 3 because it symbolized the Holy Trinity.) If numbers are to be given a mystic or occult interpretation, then the number '9' should be chosen as the *perfect* number. The reason for this is that the number 9—according to occult philosophy—is all-inclusive; any multiple of 9 can be reduced to 9 and is therefore *perfect*. For example: $9 \times 2 = 18$ $(1 + 8 = 9)$, $9 \times 3 = 27$ $(2 + 7 = 9)$, etc. up to the nth multiple.

If 9 is the 'perfect' number then there would be only one perfect meter—27/36—which consists of 9 primary units subdivided into 9 tertiary Extrametrical units for each of the primary units—$9 \times 9 = 81$ $(8 + 1 = 9)$; or thru another calculation: 3 pulses to the measure times 3 primary units times 3 tertiary Extrametrical units—$3 \times 3 \times 3 = 27$ $(2 + 7 = 9)$. QED. The '36' of the Metric Signature is explaind in the chapter on note-values. Example 33 shows the calculation of the 27/36 meter in musical terms.

EXAMPLE 33

Incidentally, this is the meter for the 5th variation of the second movement of Beethoven's Piano Sonata Op. 111, which is written as (9/16)—equivalent to (9/8). Ironically, the 27/36 meter can be notated in a more simple and practical way as 3/4—Example 34.

EXAMPLE 34

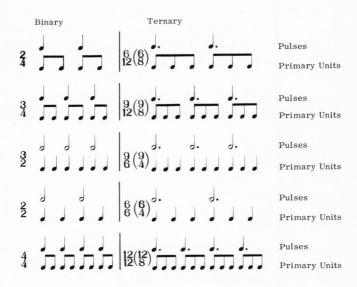

As a preliminary illustration of Binary and Ternary meters, the following paradims are presented.

EXAMPLE 35

COMBINATIVE

To the Author's knowledge, a meter that is a combination of Binary and Ternary has never been noted. This type which may be termd 'Combinative,' employs both Binary and Ternary primary units, successively or simultaneously, in the same measure, or in alternating measures.

EXAMPLE 36

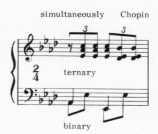

Altho only 2 examples are given there are actually innumerable specimens of Combinative meter, especially in 19th and 20th century music.*

DUPLE AND TRIPLE.

These terms, used by some theorists in place of 'Binary' and 'Ternary,' are ambiguous; they should be used only in reference to the number of pulses—not units—in a measure. In other words, traditional 6/8 is not a *Triple* meter—a meter having 3 pulses; it is a *Duple* Ternary meter—having 2 pulses (duple) with 3 primary units (ternary). Therefore, 2/8, 2/4, 2/2, 6/12 are Duple; 3/8, 3/4, 3/2, 9/12 are Triple, 4/8, 4/4, 4/2, 12/12 are Quadruple, 5/4, 15/12 are Quintuple, etc.

*The prophet of future devices, J. S. Bach, employed Combinative meter long before later composers were aware of it. Note the Metric Signature of the PRELUDE in D from the Well Tempered Clavier, Book 2. (Bach's Metric Signatures are ambiguous [and reversed] even according to traditional nomenclature. The correct signatures are in parentheses above the staff.)

TIME SIGNATURE

Time signature as a synonym for Metric Signature is imprecise. 'Time' is a synonym for 'duration,' not 'meter'—the *measurement* of duration. As the word 'tempo,' in Italian, has several musical meanings, so has the word 'time' in English: duration, pace, and rhythm.

BAR AND BARLINE

The two words are synonymous. Musicians, nevertheless, continue to define and use the word 'bar' as interchangeable with 'measure,' ignoring the truism, "Not ignorance, but the ignorance of ignorance, is the death of knowledge." Donald Tovey's explanation (*Encyclopaedia Britannica*) is definitive and clear: " . . . the bar represents no fundamental rhythmic fact. It did not come into existence so long as music was printed only in parts. When music began to be printed with all the parts ranged legibly on one page, it was necessary to score the pile of parts with vertical strokes to range them in *partitions* and guide the eye. Hence the word 'score' and the French 'partition' and German 'partitur' [and Italian 'partitura']." In short, a *measure* should never be referrd to as a *bar*.

DOTTED-HALF, DOTTED-QUARTER AND DOTTED-EIGHTH

These terms are graphic descriptions and non-defining; they do not express the durational value of the notes. As explaind in the next chapter, their durational values are as follows:

EXAMPLE 37

The same terminology must be applied for rests, as will be explaind in a later chapter.

To summarize: Meter is the grouping of pulses or units within a measure or a frame of 2 or more measures. In strictly metric music, and music based on the First, Second or Third Structure, the grouping is for 1 measure. But in music based on the Fourth or Fifth Structure, the grouping is for 2 or more measures, and comes under the general classification of Dimeter (2 measures), or Trimeter (3 measures), etc. This aspect of Meter will be fully explaind in the classification of Meters. For the moment, and to clarify the first definition of Meter, an example of a Dimeter in the Fourth Structure and one in the Fifth Structure are given.

EXAMPLE 38

A second definition of Meter is: the measurement of duration by pulses or units. In Binary, Ternary and Combinative meters, the duration is measured by pulses. For example, 2/2 is measured by 1/2 notes, 2/4 by 1/4 notes, 2/8 by 1/8 notes, etc. But in Unitary meters, 7/16, 11/16, 13/16, etc., the duration is measured by units—1/16 notes.

Meter being the measurement of duration, it has *no* relationship to Accent; Accent is neither an element nor a factor in Meter; the elements of Meter are only 3: pulses, beats and units. But Pace *is* a factor in the actual measurement of duration.

The following precise and unequivocal terminology has been presented: 'Bar' and 'Barline' are synonymous. 'Bar' is *not* synonymous with 'measure.' 'Binary' and 'Ternary' replace the traditional 'Simple' and 'Compound.' A combination of Binary and Ternary is termd 'Combinative.' 'Duple,' 'Triple,' 'Quadruple,' etc. refer to the number of *pulses*, not *units,* in a measure. 'Metric Signature' replaces 'Time Signature.' Dotted-notes and -rests are to be named according to their actual durational value, and not their graphic aspect.

QUESTIONS

1. Give two definitions of Meter.
2. What are the elements of Meter? Define each element.
3. Illustrate—as in Example 32—the following *units* in 3/4 meter: primary, secondary and tertiary, and primary extrametrical, secondary extrametrical and tertiary extrametrical.
4. Define Binary, Ternary and Combinative meter; also, Unitary meter.
5. Name 3 Binary, 3 Ternary, and 3 Unitary meters.
6. Illustrate a Combinative meter used successively; such meter used simultaneously.
7. What do the terms 'Duple,' 'Triple' and 'Quadruple' refer to? Name 2 Duple meters, 2 Triple, and 2 Quadruple.
8. What is a Metric Signature?
9. Which 2 words are *not* synonymous: (1) bar-barline or (2) bar-measure?
10. What is the durational value, in 4/4, of a dotted-half note? dotted-quarter note? dotted-eighth note?

III

NOTE—VALUES

One should not aim at being possible to understand but at being impossible to misunderstand.
—Quintilian (A.D. 35?-95?)

The 16th century rule for note-values is in full force today as then: "The semibreve [whole note] is the mother of the other notes." In modern phraseology: any note of less durational value than a whole note is a *fraction of a whole note*. To illustrate the rule, if there are 2 equal notes in a measure of 4/4, they are 1/2 notes; if 4 equal notes, they are 1/4 notes; if 8, 1/8 notes. Mathematically expresst: $2/2 = 1$ (whole); $4/4 = 1$; $8/8 = 1$. Musically demonstrated:

$$\frac{4}{4}\ \ \text{♩ ♩} = \frac{2}{2} = 1\ |\ \text{♩ ♩ ♩ ♩} = \frac{4}{4} = 1\ |\ \text{♫ ♫ ♫ ♫} = \frac{8}{8} = 1\ |$$

This method of calculation proves true also with 3/4 and 2/4.

$$\frac{3}{4}\ \ \text{♩.} = \frac{3}{4}\ \text{note}\ |\ \text{♩ ♩ ♩} = \frac{3}{4}\ |\ \text{♫ ♫ ♫} = \frac{6}{8} = \frac{3}{4}\ |$$

$$\frac{2}{4}\ \ \text{♩} = \frac{1}{2}\ \text{note}\ |\ \text{♩ ♩} = \frac{2}{4} = \frac{1}{2}\ |\ \text{♫ ♫} = \frac{4}{8} = \frac{1}{2}\ |$$

The following paradim—always presented without a Metric Signature—is familiar to all musicians and students of music.

EXAMPLE 39

35

The foregoing is the Binary system of measurement. In every book on music theory it is, despite the existence in practice of the Ternary system, the *only* one presented. It is incomplete, imprecise and unrealistic. It is incomplete and imprecise because (1) it ignores the Ternary meters, (2) there is no Metric Signature indicated, and (3) all the symbols are not shown—only note-heads, stems and beams—while there are 3 other symbols for note-values: the dot, the tie and the numeral.

The unrealistic aspect is in the disregard of rhythmic practice since the 18th century. An examination of just a small segment of the music of Western civilization will reveal the use of 1/3 notes, 1/6s, 1/12s, 1/24s, etc. In fact, almost every fraction of a whole note from 1/64 to 63/64 has been employd, not to mention such fractions as 3/20s, 5/12s and 1/28s. The misconception has been perpetuated thru loose and inaccurate terminology passt on from generation to generation, and unquestioned by students and teachers. Why should a 3/4 note be named a 'dotted-half,' and a 3/8 note, a 'dotted-quarter'? Higher mathematics is not involvd in these calculations—merely simple arithmetic.

A typical definition of a 1/4 note—in music dictionaries—is "a quarter of the *value* of a whole note," with no modifying adjective. Is the value spatial, temporal or tonal? Word dictionaries, however, are often more explicit: "A note having one-fourth the *time* [durational] value of a whole note." Such imprecision and incompleteness is not evident in explanations of *pitch* values. No teacher would ask a student to name the pitch value of ≡╪≡with no clef indicated.

It has been stated too often that Metric Signatures are not to be regarded as fractions. Why? Because with traditional terminology, the fractional factor works only with Binary meters—2/4, 3/4, 4/4—but not with Ternary meters as traditionally notated—6/8, 9/8, 12/8. The fraction in the traditionally notated Ternary meters, however, is mathematically incorrect. In order to clarify the basis of a revised notation for Ternary meters, a brief history of meters and Metric Signatures is now given.

MEDIEVAL METRIC SIGNATURES

Since the 14th century there has existed a dual metric system, in medieval terminology, (1) prolation imperfect, and (2) prolation perfect; in modern terminology, (1) binary subdivision and (2) ternary subdivision. Following is the graphic illustration.

MEDIEVAL		MODERN
C ◻ ♪ ♦♦ ♦ ⫶ ♦♩	time imperfect—prolation imperfect ꞊ 2/4	duple – binary
O ◻ ⫶ ♦♦♦ ♦ ꞊ ♩♩	time perfect—prolation imperfect ꞊ 3/4	triple – binary
C ◻ ꞊ ♦♦ ♦ ꞊ ♩♩♩	time imperfect—prolation perfect ꞊(6/8)	duple – ternary
O ◻ ꞊ ♦♦♦ ♦ ꞊ ♩♩♩	time perfect—prolation perfect ꞊(9/8)	triple – ternary

In other words, 'time' meant 'meter' according to the number of pulses, and 'prolation' meant the first subdivision (primary units) of the pulse. The number 2 was considerd 'imperfect' and the number 3, 'perfect.' The full translation of the first medieval meter is: a meter of 2 pulses (duple) with 2 units to each pulse. The first symbol of each line—the semicircle or circle—with or without the dot—is the actual Metric Signature and what follows the symbol, the interpretation of it. The circle signified 3 pulses; the semicircle, 2 pulses; the dot, 3 units to the pulse, and without the dot, 2 units to the pulse. The 'prolation' dot was adopted in later notation for the same purpose as in medieval meters, to signify 3 units to a pulse in Ternary meters. The so-calld 'dotted-quarter note' in Ternary meters is actually a 1/4 note in durational value. This will be illustrated in the modern Table of Common Note-Values.

When numerical symbols were employd as Metric Signatures, the confusion of terminology was aggravated. In the Binary meters —2/4, 3/4, 4/4—the numerator indicated the number of pulses and the denominator, the value of each pulse; for example, 4/4 meant 4 pulses of 1/4-note-value each. In the Ternary meters, however— (6/8), (9/8), (12/8)*—the numerator indicated the number of *units* (not pulses) and the denominator purportedly showed the value of those primary units; for example, (12/8) meant 12 units of 1/8-note-value each. Both 4/4 and (12/8) contain 4 pulses in a measure; but while 4/4 does equal 1, (12/8) equals 3/2. In fact, traditional (12/8) can be notated in 4/4, thereby justifying the proper signature of 12/12.

*Metric signatures in parentheses are traditional nomenclature.

EXAMPLE 40

The notational fact that is ignord is the several ways of notating a note-value in either system. For example, any one of the
following methods of indicating a whole note in Binary meter is
perfectly logical:

EXAMPLE 41

Other than note-heads, stems, flags, and beams, there are three
signs that establish note-values: (1) the dot, (2) the tie and (3)
the numeral. The numeral, especially in unusual note-values, is
precise, clearest and most reliable. With all these symbols at our
disposal there is no need to invent new symbols for note-values.
Furthermore, an invented system in music can rarely be as effec-
tive or as successful as one that has evolvd, one that has developt
from the known to the unknown, or that is the extension or clari-
fication of a principle. (A whole thesis could be written on this
very subject based on the attempts that have been, and are being,
made in proposing new symbols for note-values; but it would not
serve the purpose of this book.)
      Every irregular or unusual note-value can be calculated by a
simple mathematical formula and sometimes by a musical formula. If it is a fraction of a whole note in 4/4, either method is
practical; but if it is a fraction of any note other than a whole,
half, or quarter note, then the mathematical method is simpler. To
illustrate: if there are 5 equal notes in a measure of 4/4—♪♪♪♪♪
—each note is 1/5 × (of) 4/4 = 4/20 = 1/5. The musical
method is more involvd:

EXAMPLE 42

Actually, Example 42 is a musico-mathematical illustration. This method is used when an irregular group is for the value of 1/4 note. It is reasond thus: if there are 5 notes to a 1/4 in 4/4, there would be 20 to 4 quarters or a whole note; therefore the 5 units of the 1/4 note are 20th notes; if there are 7 notes to a quarter note in 4/4, there would be 28 notes to the whole note, and they are therefore 28th notes, etc.

When a measure is durationally less or more than a whole note, the mathematical method is simpler. Supposing there are 5 equal notes in 3/4, as in the Scriabin Etude, Op. 42 No. 1:

EXAMPLE 43

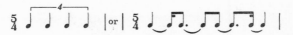

The calculation is quite simple: each of the 5 notes is $1/5 \times$ (of) $3/4 = 3/20$. It is therefore not a "quarter-note quintuplet" but a 3/20-note quintuplet. If there are 4 equal notes in 5/4, each of the 4 notes is $1/4 \times 5/4 = 5/16$ notes.

This brings up a point which proves the imprecision of traditional terminology: When is a quarter note NOT a quarter note? None of the *written* quarter notes in Examples 43 and 44 is of quarter-note value. The written quarter notes of Example 43 are 3/20 notes; of Example 44a, 1/6 notes; of 44b, 3/16 notes; and of Example 44c, 1/5 notes.

EXAMPLE 44

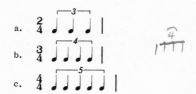

A musical diagram of Example 44b, 4 equal notes in 3/4, show three different ways of notating a 3/16 note:

EXAMPLE 45

The misconception in note-values is engenderd by the failure of composers to be specific with metric symbols, to distinguish between the graphic aspect and the actual durational value, and by the apathetic habit of allowing one symbol to function in varying ways. A classic instance of this misconception is the '256th notes' of Couperin as quoted by Nicolas Slonimsky (*The Road to Music*):

EXAMPLE 46

Because the two groups of notes have 6 beams, Slonimsky names them 256th notes. But if they are correctly notated and correctly beamd, the group of 5 notes in the first measure prove to be 32nd notes and the group of 7 notes in the second measure, 28th notes.

EXAMPLE 47

This is the calculation: on the third pulse of the first measure there are a 3/32 note (dotted-16th) and 5 equally beamd notes. Since there are 8 thirty-second notes to a quarter, and the first note is a 3/32, the remaining 5 notes must be 32nd's: 3/32 + 5/32 = 8/32 = 1/4. On the third pulse of the second measure, there are 7 equal notes to the 1/4 pulse, 1/7 × 1/4 = 1/28. If we wish to 'split hairs,' we must note that there are only 6 twenty-eighth notes sounded, the first note of the septuplet being tied to the preceding 1/2 note which, by ignoring the mordent, is a 15/28 note!

REVISED PARADIM OF COMMON NOTE-VALUES

It may seem an impossible task to re-educate professional musicians in a revised terminology. On the contrary, by adopting a system used by certain conductors in rehearsals, of referring to the *written* aspect of transposed pitches, it is quite feasible. For example, if the actual pitch should be 'G' in the part of the Horn in F, the conductor will ask for a '*written* D.' Similarly, if a musician is not familiar with the revised terminology of note-values, a 1/6-note triplet ♩♩♩ can be referrd to as a '*written*-quarter-note' triplet, and a 1/12-note triplet ♪♪♪ as a '*written*-eighth-note' triplet.

So far we have been concernd with the notation of only Binary meters, and the arithmetical calculations prove true. When we consider the Ternary meters it becomes apparent that the fractional process does not work with traditional notation. For, a measure of (12/8) does not contain 12 *eighths* but 12 *twelfths*—12/12; a measure of (9/8) contains 9 *twelfths*—9/12; and a measure of (6/8), 6 *twelfths*—6/12.

With a precise nomenclature for Ternary meters there is no
need to distinguish between the 'augmentation' dot and the
'ternary' dot—they are actually one and the same. The 'augmentation' dot increases the value of a note by 1/2; in 4/4, a 1/4 note
with a dot ♩. is 1/4 + 1/8 = 3/8. The 'ternary' dot purportedly indicates that the pulse contains 3 primary units: (12/8)
♩. = ♫ But if the Metric Signature is correct, the dot is still an

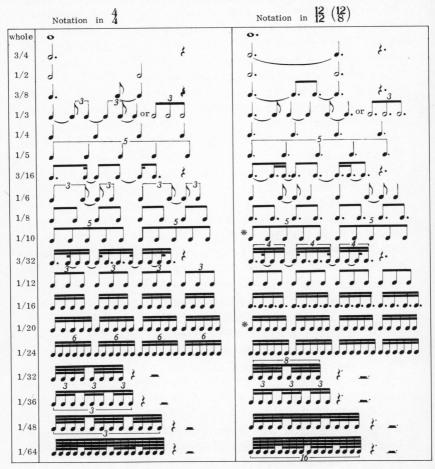

*The numeral is sufficient with this note-value, and the dot is eliminated.

'augmentation' dot. For example, if a 4-pulse Ternary meter is notated as 12/12, the 'written quarter note' is actually a 1/6 note and the dotted-quarter ♩. is $1/6 + 1/12 = 3/12 = 1/4$ note. The revised Table of Common Note-Values (q.v.) on page 42 graphically illustrates this.

As mentioned earlier there are, besides the 'dot,' two other symbols for indicating note-values: the 'tie' and the 'numeral.' The tie usually functions when a note overlaps a pulse, unit or a bar—see 1/3 and 3/32 notes in the Table of Common Note-Values. The numeral functions in irregular groups such as the triplet ♪♪♪ in Binary meters, and in groups which are irregular in any meter such as quintuplets and septuplets.

With these important points regarding note-values, we are ready to construct a revised Table of Common Note-Values. Consistent with the arguments presented, the Table is arranged in 2 columns: the left column for Binary notation, and the right for Ternary notation.

Special note must be made of the difference in notation between binary and ternary meters. With all note-values, the ternary meters require the dot; for the sake of convenience, however, 1/10-, 1/20-, and 1/32-notes omit the dot and substitute the numeral alone. In binary meters, ternary note-values—1/3s, 1/6s, 1/12s, etc.—require the tie and/or the numeral. In ternary meters, binary note-values—1/2s, 1/4s, 1/8s, etc.—require only the dot; and 3/8s, 3/16s and 3/32s require the dot and the tie. Note-values that are irregular in both meters—quintuplets, septuplet, etc.—require only the numeral.

The alternate notation for 1/3 notes ♩♩♩ is explained. In rhythms based on 'cross-accents'—3 beats against 4 pulses, as in 1/3 notes; 4 beats against 3 pulses, as in 3/16 notes; etc.—and in complex irregular rhythms such as 4 irregular beats against 4 pulses, there is a 'slow' and a 'fast' notation. In 'slow' notation, the pulses must be evident and clearly shown for the rhythm to be felt. To illustrate: with 3 beats and 4 pulses at a slow pace, the first notation of 1/3 notes is practical of execution since the exact point at which each pulse occurs is clearly shown. But at a fast pace, in which the entire *measure* becomes the pulse, the

contradiction: 1 measure / 1 pulse

alternate notation $\overset{3}{(\text{♩♩♩})}$ is quite practical. The fast notation is simpler in *appearance* than the slow notation, but not in execution. To be convinced, try playing 3 beats against 4 pulses at ♩ = 60, reading the 'fast' notation, and then play the same rhythm at a pace of ♩ = 160; or reverse the process, and play the rhythm at a slow pace twice, reading first the 'fast' notation and then the 'slow' notation.

The disregard by many contemporary composers of practicality of notation in complex rhythms has aggravated the confusion in both rhythmic and metric notation. It cannot, therefore, be too strongly stresst that, in irregular and complex rhythms, the usual simplistic notation is practical only in *fast* pace but difficult in *slow* pace. Hence, several other examples are now presented.

In such rhythms as 3 beats in 2/4, 4 beats in 3/4, and 3 beats in 5/4, if the pace is sufficiently fast to feel the *measures* as pulses, then *fast* notation is practical.

EXAMPLE 48

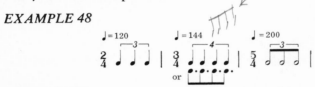

If, however, the pace is slow, or if the group overlaps—at any pace—the bar, then the notation must be such that each pulse can be felt.

EXAMPLE 49

It is a mystery why so many contemporary composers use such impractical notations as in Example 49b; it is an unnecessary obstacle placed before the performer.

In irregular groups that require a numeral, the *written* value of the units must be determind. Example 48 presents this problem: the 1/6 notes in 2/4, and the 3/16 notes in 3/4, are written as 1/4 notes. The rule for such cases is: if the number of *irregular* units occurs between 2 *regular* units, the written value of the *larger* unit is used. Applying this rule to Example 48, the problem is stated thus: in 2/4 we want 3 notes for the entire measure or a half note, there are 2 quarter notes and 4 eighth notes to the measure, the 3 notes come between the 2 and 4, therefore the 3 notes should be written as quarter notes (the larger of the 2 regular units). By graphically illustrating the calculation in the 3/4 problem, the rule may be more clearly explained.

Number of irregular units in 3/4— ♪♪♪♪ (4) for a ♩.
There are 3 quarter notes ♩ ♩ ♩ and 6 eighth notes ♫♫♫,
The 4 notes are between the 3 quarters and the 6 eighths.
Therefore the 4 irregular units must be written as ♩♩♩♩, quarter notes plus the numeral.

If the reader refers to Example 47, he can see why the irregular group of Couperin's so-calld '256th notes' in the second measure were written as 16th notes. There are 7 notes to a quarter note in 3/4, which is irregular; there are, regularly, 4 *sixteenths* or 8 *thirty-seconds;* the 7 notes come between 4 and 8; therefore they are written as *sixteenths,* the larger unit.

Note that in the revised Table of Common Note-Values, no unit smaller than a 64th note is present, altho there are rare cases in which 128th notes are used. If such a note-value were required it is a simple matter to use a larger denominator in the meter. For example, if a metric signature of 3/4 would result in 128th notes—with 5 beams—changing the metric signature to 3/2 makes the 128th notes 64th notes.

The problem of properly alining irregular groups is not difficult. Taking as an illustration Example 49a—the 3/4 and 5/4 examples—the following is the method.
PROBLEM: Aline 4 equal notes in 3/4. There are 3 pulses and 4 beats. Find the common denominator of 3 and 4, which is 12. Divide that number (12) by the number of pulses (3), which is

four. That means we must have 4 units to each pulse (1/4-note-
value) = 4 sixteenth notes. Write the 3 groups of sixteenths
with *stems* only. Bracket every 3 sixteenths. Aline each *beat* with
the first stem of each group.

EXAMPLE 50

The second rhythm—3 equal notes in 5/4—is diagramd as
follows:

EXAMPLE 51

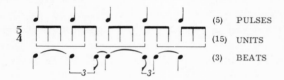

This method is a great time and work saver when many units are
involvd, as in the example from Chopin's Ballade in F minor, Example 24f—Chapter I.

In the revised table of Common Note-Values, note that the
'written' half notes to designate 1/3 notes are *beamd* instead of
bracketed. Logically, written 1/2 or 1/4 notes that require a
numeral use the bracket to preclude ambiguity; but there can be
no ambiguity with 1/2 notes since the note-heads are not black.
Besides, a numeral with a beam means a particular number of
equal units in the pulse, so the white notes with a beam could not
be mistaken for 8th notes or 12th notes.

The quotation of single notes is not shown for it is a simple
matter to substitute the same number of flags for beams: 1 flag
for an 8th note, 2 for a 16th, etc. In contemporary practice, however, the flag is gradually disappearing. The substitution is accomplisht by the use of the *extended beam* and the *short stem*.

EXAMPLE 52

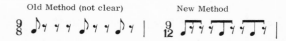

The new method, however, should be used only in complex rhythmic patterns involving many rests.

EXAMPLE 53

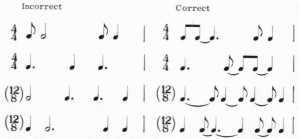

As mentiond earlier, there are several ways of symbolizing note-value. Example 41 shows 4 ways of notating a whole note. Nevertheless, students who are not cognizant of the several symbols for note-value are often guilty of the most confusing notation even when no beams are involvd. Example 54 shows, in the first column, some of the errors of notation and, in the second column, the correct notation.

EXAMPLE 54

In each *incorrect* example, the sum of the note-values does equal the entire measure; but it is not *measured* by pulses and units. The rule to observe is: Note-values are distributed in a measure according to pulses and/or units. In other words, a single symbol should not indicate a fraction of one pulse plus a fraction of another. For example, the half note in the first 'incorrect' example contains an 8th of the first pulse and 3/8 of the second and third pulses; in the second 'incorrect' example, the quarter note contains an 8th of the second pulse and an 8th of the third pulse, etc.

Even the masters have been guilty of misnotation, particularly in ternary meters (Example 55.) But the misnotation can be justified for several reasons: the accompanying part is properly alined to show the exact duration of each note in the melody, the rhythm rather than the meter is shown, and the piece is for a keyboard instrument.

EXAMPLE 55

The third reason that was given as justification for the notation of the melodic part can be further clarified. The melody is in the First Structure rhythm, 3 beats against 2 pulses, in other words, 3/4 against 6/8, in both examples above. But the basic meter of the piece is (6/8), and the notation for all parts should conform to that meter. This point will be more fully considerd in the Classification of Meters.

DOUBLE-DOT AND TRIPLE-DOT

The use of the Double-Dot and the Triple-Dot is a convenient abbreviation, but can prove to be impractical; they should be used only in *fast* notation and in common rhythmic patterns. The Triple-Dot should be used only in certain irregular note-values and in music—fast or slow—for keyboard instruments. Otherwise, both dotted symbols should be replaced by the tie.

*EXAMPLE 56

*Single and Double-Dotted Notes, and Single-Dotted and Tied Notes are symbols of some very common note-values. In Example 56, the note-values are:

♩.. = 7/8 note. *Calculated*: 1/2 (4/8) + 1/4 (2/8) + 1/8 = 7/8

♩.. = 7/16 note. *Calculated*: 1/4 (4/16) + 1/8 (2/16) + 1/16 = 7/16

♪. = 7/32 note. *Calculated*: 1/8 (4/32) + 1/16 (2/32) + 1/32 = 7/32

Lower line is calculated in the same manner as the upper line except that a tie is used in place of the second dot.

Note that the 15/8 is *not* the traditional ternary meter, but a *Unitary* meter similar to the 15/16.

The durational value of notes in a *rubato* passage or in a *fermata*—long or brief—is not considerd. Such durational values are too indefinite, vary with each performer, and could be measured only by scientific instruments. Furthermore, the total duration of the entire rubato passage, to be musically correct, must be the same as if the passage would be playd at a strict pace.

UNUSUAL NOTE-VALUES

In 19th- and 20-century music are found many unusual note-values, but theoretically, the possibilities are virtually numberless. To tabulate either the actualities or the possibilities would be a monumental and unnecessary project. However, in order to lend some support to the argument that "almost every fraction of a whole note has been employd," several unusual note-values that are found in only five pieces by one composer—Chopin—are analyzed.

Example 57a.   (*) 11 notes to a 3/4 note: 1/11 × 3/4 = 3/44 notes.

Example 57b.   (*) 7 notes to a 3/4 note: 1/7 × 3/4 = 3/28 notes.

Example 57c.   (*) 5 notes to a 1/8 note: 1/5 × 1/8 = 1/40 notes.

Example 57d.   (*) 7 notes to a ternary 1/4 note: 1/7 × 1/4 = 1/28 notes.
              (**) 14 notes to a ternary 1/4 note: 1/14 × 1/4 = 1/56 notes.

Example 57e.   (*) 5 notes to a 1/8 note: 1/5 × 1/8 = 1/40 notes.
              (**) 3 rather common note-values: 7/16 note, 1/48 notes, 1/24 note.

Example 57f.   (*) 6 notes to a 1/12 note: 1/6 × 1/12 = 1/72 notes.

Example 57g.   (*) 24 notes to a ternary 1/4 note: 1/24 × 1/4 = 1/96 notes.

EXAMPLE 57

An explanation of the metric signature of Example 57c is in order. Note that the 12/8 is not the *traditional* 12/8 meter; the *8th note* is the *pulse* and it is therefore a true 12/8 and not 12/12. This point will be further clarified in the Classification of Meters.

QUESTIONS AND EXERCISES

1. Illustrate 4 ways of indicating—in binary meter—a whole note; 3 ways of indicating a 3/8 note and a 3/16 note.
2. Illustrate the various ways of the same note-values—whole, 3/8, and 3/16—in ternary meter.
3. Illustrate the notation—in both binary and ternary meter—of a whole note, 3/4 note, 1/2 note, 1/4 note, 1/8 note and 1/16 note.

4. What is the durational value of a 'dotted-half,' 'dotted-quarter,' 'dotted-eighth' and 'dotted-sixteenth,' in binary meter? What is durational value of the same notes in ternary meter?

5. Calculate the durational value of the following 'written' quarter notes:

6. Calculate the note-value of the 3 equal notes in 5/4 of Example 48.

7. Correct the following notational errors:

8. Illustrate the following *unusual* note-values in binary meter:

1/40 notes   3/40 notes   1/72 notes   1/96 notes   1/28 notes

IV

RESTS, BEAMS, FLAGS

The traditional rules regarding rests—the durational value of silences—are simple, and the very few revisions suggested simplify them further. Yet, the common errors in rest-notation that even advanced students of theory and composition make are most discouraging. Following are the rules regarding Rests.

RULE 1. The durational value of any rest, except the whole rest, is equivalent to the *note-value.* (Note the *exact* placement of each rest.)

EXAMPLE 58

A whole rest means silence for the *entire measure,* whether the meter is as small as 2/16 or as large as 12/4; a whole rest means the same for binary and ternary meters and is always placed in the *center* of the measure.

RULE 2. Traditionally, in ternary meters, a rest for 1 pulse is written as 1/4 + 1/8 rest: ⸹ ⸹ . A revised method is perfectly logical, which is to add a dot to the 1/4 rest, as is done with the note: ⸹· . (All references to rests are as the *written* aspect, not the actual durational value. The written quarter rest in ternary meters is actually 1/6 in durational value.) The dot may be added

also—in ternary meters only—to the 1/2, the 1/8 and the 1/16 rest (Example 59a); it should never be added to the whole rest. (Example 59b). In binary meters, only the dotted 1/8 rest should be used (Example 59c).

EXAMPLE 59

RULE 3. In 4/4 and 12/12, the 1/2 rest (dotted in 12/12) is used only on the 1st or 3rd pulse, never on the 2nd (Example 60a); 1/4 rests must be used on the 2nd and 3rd pulses. In 3/4 and 9/12, the 1/2 rest is never used—two 1/4 rests are used instead (Example 60b). In true 'compound' meters such as 5/4 and 15/12 (traditional 15/8), which are generally treated as 3/4 + 2/4 or 2/4 + 3/4 in 5/4, and as 9/12 + 6/12 or 6/12 + 9/12 in 15/12, the rests must be assigned according to the particular subdivision (Example 60c).

EXAMPLE 60

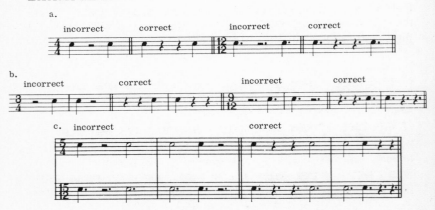

RULE 4. In ternary meters, the written 1/4 rest (𝄽) is not used at the 2nd primary unit; 2 written 1/8 rests are used instead. But the 1/4 rest may be used at the 1st of each group of primary units.

EXAMPLE 61

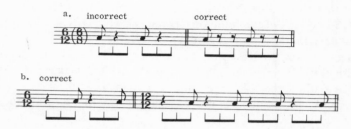

RULE 5. A 1/4 rest should not overlap the last unit of one pulse to the first unit of the next.

EXAMPLE 62

Rule 5 applies also to secondary, tertiary and quartan units.

EXAMPLE 63

The rules given are for binary meters with 4 as denominator—
4/4, 3/4, 2/4—and ternary meters with 12 (traditionally 8) as
denominator—12/12 (12/8), 9/12 (9/8), 6/12 (6/8). As will
be shown later, 4/4, 4/2, and 4/8 are relatively similar in measurement, the only difference being the value of the *pulse*. Traditionally, 12/8, 12/4, and 12/16 are also similar with a difference in
the value of the *unit*. It is not difficult to apply the rules for 3/4 to
3/2, 3/8 or any of the common meters.

The remark earlier regarding note-values smaller than 1/64
applies as well to Rests, namely, that if a 3/4 meter requires a
128th rest, changing the metric signature to 3/2 makes the 128th
rest a 64th. Rests of longer duration than a whole measure are
not considerd, as there is no confusion in traditional practice.

EXAMPLE 64

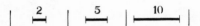

Not considerd also are unspecific silences such as breath-pause
₍,₎   at the end of a phrase, and 'breaks' (*//*).

EXERCISES

Correct the following errors in the notation of rests:

BEAMS

An explanation of the terminology regarding beams and flags will simplify comprehension of the rules formulated:

Primary Beam—A single beam connecting primary units.

Secondary Beam—Added to the Primary Beam and forming a Double Beam.

Tertiary Beam—Added to the Double Beam and forming a Triple Beam.

Quartan Beam—Added to the Triple Beam and forming a Quadruple Beam.

Short Beam—Analogous to the leger line in pitch notation; if a
secondary, tertiary or quartan beam is necessary for a single
note, it is shortend.

Extended Beam—The modern practice of extending a beam or
beams in place of a flag or flags when all the units of a group
are not written.

Short Stem—Used in place of the omitted note of an extended
beam.

Broken Bracket—Used in irregular groups requiring a numeral
with unbeamd notes. The bracket or broken bracket is preferable to the slur as it is 'impossible to misunderstand' it.
In irregular groups of written 1/2 notes, the beams may be used as
there is no ambiguity.

Extrametrical—Units or groups of units not implied in the metric
signature. For example, a triplet in Binary meter or a duplet
in ternary meter is an *extrametrical* group of units.

Of the following rules, some are revised because the traditional rules are irrational and ambiguous, and some are retaind
because they are 'right.'

RULE 1. Beams should cover an entire pulse or unit-group and
not overlap to a fraction of another pulse or unit-group. In 2/4

and 4/4, however, the Primary Beam may cover 2 pulses in the same measure.

EXAMPLE 65

RULE 2. Single units (primary, secondary, et al.) within a pulse should be notated either with a flag or extended beam, and not be included with other beamd notes.

EXAMPLE 66

RULE 3. Double, triple, and quadruple beams should be broken according to the secondary units for the double beam, tertiary units for the triple beam, and quartan units for the quadruple beam.

EXAMPLE 67

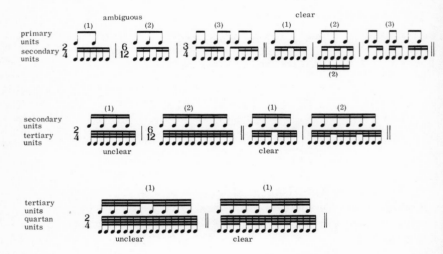

A distinction often ignord is that between a sextuplet and a double-triplet. The 6 notes in 2/4 (Example 67—secondary units) are not a sextuplet but a double-triplet, as shown in the 'clear' version. Hence, if a metrical subdivision is intended, Rule 3 should be followed. There are, however, instances when a composer intends the extrametrical pattern and indicates it by phrasing or accentuation (Example 68).

EXAMPLE 68

Sextuplet   Double-Triplet

The sensitive performer distinguishes between the sextuplet and the double-triplet, but the composer must make his intentions clear in the notation.

If the beams are correctly drawn, the numeral can be eliminated in any group up to 6 units. For example, if the 3 + 3 + 2 rhythm is beamd as in Example 69a, it might be interpreted as in 69b; it should be beamd as in 69c. And the three eighth notes in 69d might be interpreted as a triplet (69e); it should be notated as in 69f. But if Example 69b were in 3/4, the beaming is correct and the numeral can be eliminated.

EXAMPLE 69

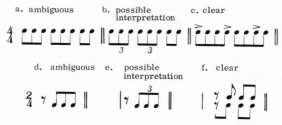

To most students, correct beaming is a deep mystery. There is some excuse for the confusion when extrametrical rhythms are involvd, when, for instance, in (6/8) the hemiola rhythm is notated as 3/4.

EXAMPLE 70

If the metric pulses and units are kept constantly in mind and the beams are contain within each pulse and unit—not overlapping—there is no misunderstanding.

EXAMPLE 71

Units

Pulses

RULE 4. Short beams must be written *within* the units to which they belong. This rule is difficult to express verbally; graphic illustration is best.

EXAMPLE 72

Note that the (6/8) rhythm is marked 'slow.' If, however, the pace is *fast* and the extrametrical rhythm is intended, the "incorrect' notation is permissible, as Example 73 illustrates.

EXAMPLE 73

RULE 5. Overlapping rhythms that involve beams should be notated *metrically—*not *rhythmically—*correct, especially in ensemble music. In solo and keyboard music, the rule is not as important, but in ensemble music the pulses of each measure must be clearly evident.

EXAMPLE 74

Overlapping beams are quite prevalent in modern practice, but the 'innovation' is not an improvement over the traditional method. Such notation of extrametrical rhythms creates a needless obstacle for the performer. Furthermore, composers who use overlapping beams in conjunction with changing meters contradict their conception of the barline. The following example of contradiction—from a contemporary work—is typical.

EXAMPLE 75

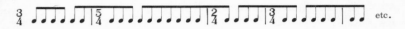

In all matters of notation, Quintilian's axiom must be constantly borne in mind: "One should not aim at being *possible* to *understand* but at being *impossible* to *misunderstand*" [Author's italics].

EXERCISES

Correct the following errors in notation of *beams*:

1.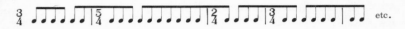

2.

3.

4.

5.

6.

7.

8.

9.

10.

FLAGS

There is only one rule regarding flags: a flag is used *only* with a single note within one pulse. In durational value, a single flag is equal to a single beam, a double flag, to a double beam, etc. In pre-20th century vocal music, the rule for flags was not applied to syllabic setting of words in beamd patterns; but in 20th-century vocal music it is the general rule. For even in a rhythmic pattern as simple as in Handel's "Hallelujah" (Example 76), the new notation is more practical.

EXAMPLE 76

Sometimes the rule was applied to instrumental music, purportedly to clarify the phrasing. Compare, however, the old notation with the new in the following examples and note the clearness of the modern method.

EXAMPLE 77

As stated earlier, the flag is gradually disappearing in contemporary practice, sometimes to an irrational degree. Either the short stem and extended beam should be used—especially when rests are involved (Example 78a), or an overlapping note is divided in 2 and a tie connects the two notes (Example 78b).

EXAMPLE 78

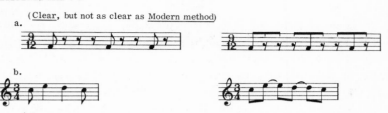

One of the most confusing examples of impractical flagging is found in the Nr. 5 Zeitmasse of Stockhausen. (Example 79). Only the rhythmic values are shown.

EXAMPLE 79

Because of the irrational use of flags and rests (overlapping and ternary rests in a Binary meter), neither the individual part nor the score is easily decipherable. Such amateurish notation places an unpardonable obstacle in the path of the performer. Example 80 shows a rational notation. The 16th stems at the bottom are for proper alinement.

EXAMPLE 80

The foregoing is an example of complicated notation for a rather simple rhythm. The next example—from the same composer —is probably the extreme of irrationality (81a). The calculation is shown in 81b.

EXAMPLE 81

It required a half hour to determine the calculation of the irregular groups. Even with a logical and clear notation, the demands made on the performer's rhythmic ability are unconscionable.

It is understood that a single flag is equivalent to a single beam, a double flag, to a double beam, etc. Example 80 (q.v.) shows the substitution of double beams for the double flags that are in Example 79.

EXERCISES

Clarify the following notations by either retaining single flags, or substituting beams or ties wherever practical:

1. cum sancto Spi - ri - tu in glo-ri-a De-i   Pa-tris.

2. per quem o ——————— mni - a fa - cta, fa - cta sunt.

V

METERS AND METRIC SIGNATURES

Statistics are no substitute for judgment.
—HENRY CLAY

Chapter 10 of Gardner Read's book *Music Notation*, entitled "Meter and Time Signatures," begins: "TIME, METER, TEMPO, RHYTHM—musicians often use these terms imprecisely. But the four expressions are not synonymous, and only two are interchangeable. The analogous terms are *time* and *meter*. . . ." The first sentence and the first phrase of the second sentence state a truism; the remainder of the statement vitiates the truth.

The more than two dozen dictionary definitions of 'time' can be condensed into one synonym, 'duration,' which is present in every definition of 'time.' The word 'meter' is derived from the Greek *metron*, the Latin *metrum,* and the French *mètre;* in ANY language, it means 'measure.' In music, therefore, 'Time' means *'duration'* and 'Meter' means *'measurement of duration'*. Hence, 'Time' and 'Meter' are NOT analogous or interchangeable.

In Italian, 'tempo' has *four* different musical meanings: (1) pulse, (2) meter, (3) movement (of a Sonata, Suite, etc.), and (4) pace. What American musicians call 'tempo' is termd by Italians *'andamento'*—meaning 'pace'—which is the fourth definition of 'tempo' given above. In the Italian equivalent of the word 'meter', Mr. Read gives *'metro'*. That term is applied only to *poetic* meter; the correct translation of 'meter' is 'tempo'—the second musical meaning of the Italian word.

With the imprecise terminology of meter and its elements that has been in use for about 400 years, and which was dealt with in chapter II, it is not surprising that some contemporary meters and metric signatures are in a state of utter confusion. The first false step is taken when a rhythmic structure replaces the metric signature or when the beaming of units is ambiguous. Example 82 illustrates both fallacies.

EXAMPLE 82

The metric signature in each example shows the second rhythmic structure—irregular subdivision, which is NOT a meter, and the beams are rhythmically correct but metrically ambiguous. The irregular subdivision of 4 + 2 + 3 is derived from 9/8, which is the correct traditional signature; and the beams should be as in Example 83a. The 3 + 3 + 2 rhythm (Example 82b), also irregular subdivision, is derived from 4/4 meter, and should be notated as in Example 83b.

EXAMPLE 83

A case of *reductio ad absurdum* in 'rhythmic,' in place of 'metric,' signatures is the "probably unsurpassed signature" of a piece by Daniel Jones (quoted in Curt Sachs's, *Rhythm and Tempo*: 3 + 2 + 3 + 2 + 2 + 3 + 2 + 2 + 2 + 3 + 2 + 2 + 3 + 2 + 3 + 3 + 2 + 3 + 3 + 3 + 2 + 2 + 3 + 2 quarters. The series adds to 59 quarters; by adding a quarter rest in the last measure, making 60 quarters, we obtain 15 measures of 4/4, a sensible metric signature.

The type of metric signature that is in reality a rhythmic notation is termd by one writer 'compound meter'; and he has in his Table of Compound Meters such illogical arrangements as:

$$\frac{4}{8} = \frac{1+3}{8}, \quad \frac{6}{8} = \frac{1+2+3}{8}, \quad \frac{4}{4} = \frac{1+2+5}{8}, \quad \text{and} \quad \frac{9}{8} = \frac{1+2+6}{8}.$$

The fallacy of Bartok's irregular subdivision signature (Example 82) is COMPOUNDED in the 'compound meters' by the presence of the number 1 as the numerator. It must be rememberd that Meter is the *measurement* of duration—not simple duration; that in measuring duration—as well as distance—2 points are necessary. To analogize: let us imagine that the dial of a clock represents a whole note, that the clock has an hour-, a minute-, and a second-pointer or hand, and that this clock has no numbers or dots to indicate the hours, minutes and seconds. A complete revolution of the second-hand would indicate the *duration* of one minute; it would *NOT* show the *measurement* of that duration. Hence a metric signature of 1/1 (one whole note) would merely indicate the *duration*, not the *measurement,* of the duration. Therefore, in music, a measure must have at least 2 pulses, a rhythm must have at least 2 beats, and a pulse or beat must have at least 2 units. The irregular subdivision of Bartok's Mikrokosmos #153 is logically 3 + 3 + 2, −3 irregular beats against 4 pulses derived from 8 units. If the 8 units had been grouped into 1 + 4 + 3, the first beat would be *unmeasured,* and the rhythmic signature would be illogical.

Another illogical metric signature is that which has a numerator of 1 among multimeters, such as 1/1, 1/2, 1/4, 1/8, 1/16, and even 1/32. (Multimeters have also been termd 'variable,' 'changing,' and 'mixed,' and the statistics of these single-pulse or single-unit meters can be found in Gardner Read's *Music Notation*). This malpractice results in a second fallacy: the separation of a single pulse or unit from the preceding or following measure to which it rightly belongs.

To clarify the argument, let us examine the three measures that have a numerator of 1, in the second of Stockhausen's "nr. 2 Klavierstücke I-IV." (Only the rhythmic values are given in Example 84.)

EXAMPLE 84

Measure 16   Measure 25   Measure 31

The two measures preceding and the two following the 1/4 measure (16) form this sequence: 5/8 - 3/8 - 1/4 - 3/8 - 2/8, which equal 15/8. The five measures of changing meters can be converted to three measures of 5/8 by combining the second and third measures—3/8 + 1/4 (2/8) = 5/8—and by combining the fourth and fifth measures—3/8 + 2/8 = 5/8. Measure 25 is preceded by a 4/8 measure and followed by a 5/8 measure; the sequence is: 4/8 - 1/8 - 5/8; the first two measures can be combined to form one measure of 5/8, 4/8 + 1/8 = 5/8. Measure 31 is preceded by four measures of 5/32 which equal 20/32 or 5/8. In all three cases multimetric signatures were unnecessary and the particular segments could have been notated in 5/8.

Such illogical notation as just analyzed is the result of certain contemporary composers' misconception of the barline's function. As explaind in Chapter II, "the barline represents no fundamental rhythmic fact"; the first pulse or unit of a measure is NOT necessarily strong, accented or even sounded; the barline is for *metrical,* not *rhythmic,* subdivision; and multimeters stem from the rhythmic principle of Overlapping—the Third Structure. In Chapter Eight of Creston's *Principles of Rhythm,* the multimeters of compositions by Stravinsky, Bartok, Prokofieff, et al. are converted to monometric—single metric—signature thru Overlapping. Since we are concernd at present with Stockhausen's composition, let us see if that can be so converted.

Stockhausen's piece has a change of metric signature at almost every measure, and employs denominators of 4, 8, 16 and 32. The first seven signatures are: 2/8 - 3/8 - 4/8 - 6/32 - 4/8 - 1/8 - 5/8. The smallest denominator is the thirty-second note. Converting the larger denominators to thirty-seconds we obtain 382 thirty-seconds for the entire piece. Reconverting the thirty-seconds to eighths—by dividing 382 by 4—we obtain 95 eighths plus a sixteenth. Since the last measure of the piece has a fermata, we can increase the sixteenth to an eighth, and make the total number of eighths, 96—which number can be divided by 3—and give us 32 measures of 3/8.

The argument regarding multimetric versus monometric notation is not to be construed as the forbiddnace of any change whatsoever in metric signatures. An occasional change can be logical, but continual changes—if the composer's rhythmic sense is impeccable—imply a basic common denominator. The reader should re-examine Example 22, Chapter I, and note how the 16th-century composer achieved multimeters with a single metric signature thru Overlapping. Then the reader should examine specimens of multimetric, converted to monometric, notation in several 20th-century works (Chapter Eight of *Principles of Rhythm*).

The First Rhythmic Structure—Regular Subdivision—has also been used as a metric signature, often unnecessarily or ambiguously. If the hemiola (3 pulses vs. 2 beats, or 2 pulses vs. 3 beats) is the rhythmic basis, it is unnecessary to indicate it in the metric signature by 3/4 (6/8) or 6/8 (3/4). Example 85 illlustrates the unnecessary and the correct signatures for these.

EXAMPLE 85

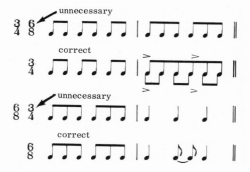

A specimen of ambiguous notation of Regular Subdivision is Ravel's *Chanson Romantique* from "Don Quichotte à Dulcinée" (Example 86).

EXAMPLE 86

In Ravel's notation there is no indication of whether the 3/4 measure means ♩=♩. or ♪=♪ of the preceding measure; the correct metric notation, therefore, is as in Example 87, eliminating the change of meter.

EXAMPLE 87

Another unnecessary change is the indication of Combinative meter—also termd 'Combined' meter—in the signature (Example 88).

EXAMPLE 88

Combinative Meter is actually a form of the first Rhythmic Structure, as can be illustrated by regular subdivisions other than the hemiola (Example 89).

EXAMPLE 89

Next we consider Sequential meters—also termd 'Alternating' meters—as found in 20th- as well as pre-20th-century music. Example 90 shows several specimens.

EXAMPLE 90

> 3/4:4/4—Brahms, *Variations on a Hungarian Song*
> 5/8:4/8—Scriabine, Prelude, Op. 11 No. 16
> 3/4:4/4—Richard Strauss, *Der Rosenkavalier*
> 6/4:4/4—Liszt, *A Faust Symphony*
> 4/4:5/4—Henry Cowell, *Exultation*

As long as only two metric signatures are involvd, these could be termd 'alternating' meters. But in order to include three or more signatures, as in Example 91, the more appropriate term would be 'Sequential.'

EXAMPLE 91

> 9/8:6/8:8/8—Creston, *Pastorale*, Suite for Saxophone and Piano
> 4/4:3/4:4/4—Creston, String Quartet, First Movement

When only two or three metric signatures are used—which do not add up to an unwieldy large meter—it is more practical to either combine the meters into one, or use the smallest meter with overlapping. To illustrate: the Scriabine *Prelude* (Example 90), with the metric signature 5/8 4/8, totals 9/8; the 9/8 signature would be truly metric and the measure would be based on the second rhythmic structure—irregular subdivision. The Creston *Pastorale* (Example 91), 9:6:9/8:8:8 totals 24/8; this can be converted to 4 measures of 6/8, and would be based on the fifth rhythmic structure—irregular subdivision overlapping. Example 92 illustrates the two specimens musically.

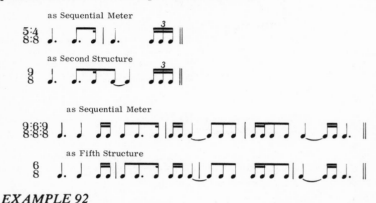

EXAMPLE 92

Believing that the distinction between binary and ternary meter would be clarified by his innovation, Dalcroze substituted the written quarter note for binary, and the written dotted-quarter note for ternary meter as the denominator (Example 93).

EXAMPLE 93

As was pointed out earlier, however, any note symbol without a metric signature is metrically indeterminate and meaningless. Villa-Lobos, on the other hand, eliminates the denominator altogether in some of his metric signatures (Example 94). This practice would not be impractical if binary or ternary meters were in force exclusively. But if there should be combinative or unitary meters employd, it would be imprecise.

EXAMPLE 94

Perhaps the most illogical and confusing type of meter is the so-called 'fractional' or 'decimal' meter. This type ignores all principles of rational and practical notation, impedes interpretation and performance, and complicates a simple and traditional rhythmic practice. Several of these 'fractional' meters are unriddld without further comment (Example 95).

EXAMPLE 95

a.

rational notation

= Irregular subdivision

b.

rational notation

= Irregular subdivision

c.

$3\frac{1}{2} = \frac{3}{4} + \frac{1}{8}$ ♩ ♩ ♩ ♪

rational notation

$\frac{7}{8}$ ♩ ♩ ♩ ♪   = true Compound meter or Unitary
   $\frac{4}{8}$ + $\frac{3}{8}$

d.

$2\frac{1}{2} = \frac{2}{8} + \frac{1}{16}$ = ♩ ♩ ♩

rational notation

$\frac{5}{16}$ ♩ ♩ ♩   = Unitary meter

Certain uncommon metric signatures, altho logical, are un-necessarily complex and rather impractical. Example 33, chapter II (q.v.) demonstrates a metric signature of 27/36, which is precisely calculated to the smallest unit employd. The simple and practical notation is shown as traditional 9/8 in Example 33, and as 3/4 in Example 34.

In Scriabine's Prelude, Op. 31 No. 3 (Example 96), the 10th notes of the upper part are the smallest units employd; the metric signature could be 10/10 for the upper part and 2/2 for the lower part. Wisely, Scriabine notated it simply as 2/2 for both parts, since it is a kind of combinative meter employing 10th-note units in one part and half-note pulses in the other.

EXAMPLE 96

$\frac{2}{2}$ ♫♫♫♫♫ ♫♫♫♫♫

$\frac{2}{2}$ ♩   ♩ ♩

The opening section of Creston's *Dance Overture* indicates both the precise-(traditional)-but-complex notation—18/16—and the simple-(traditional)-and-practical notation—9/8 (Example 97).

EXAMPLE 97

$\frac{18}{16}\left(\frac{9}{8}\right)$ ♩.♫ ♩.♫ ♫♫♫ |

Other examples—all from Scriabine—of this type of combina-tive meter are:

EXAMPLE 98

To find the common meter in a combinative meter, the various fractions must be reduced to a common fraction. The divisor in each case is the number of primary units and extrametrical units, to the pulse. To illustrate: Analyzing the 'Complex-Precise' notation of Example 98b, there are 3 primary units and 5 extrametrical units to the pulse. Dividing the 9/12 by 3, we obtain 3/4; and dividing the 15/20 by 5, we also obtain 3/4. Example 98a is calculated the same way, except that the 20th notes are not as obvious because the pattern shows the number of *beats* and not the units.

In Example 98c there are 3 primary units and 9 extrametrical units to the pulse. Dividing the 6/12 by 3 (primary units), we obtain 2/4; and dividing the 18/36 by 9, we obtain again 2/4. However, since 3 primary units indicate ternary meter, we must reconvert the binary 2/4 to 6/12, the ternary equivalent. Otherwise it would be a more complex notation as Example 99 shows.

EXAMPLE 99

Debussy's fifth Prelude, *Les collines d'Anacapri*, is notated as
the Combinative meter, 12/16 + 2/4, whereas the meters employd in the entire piece are (12/16), 2/4, (6/8), and (3/4).
All four meters can be notated in 2/4. (Metric signatures in parentheses are the traditional nomenclature.)

EXAMPLE 100

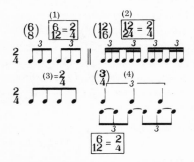

A case in which an unusual metric subdivision is in both parts
is No. 89 in Book 5 of Creston's *Rhythmicon* (Example 101). In
that piece, the pulses are subdivided in both the melody and the
accompanying figuration NOT in duplets, triplets, quadruplets or
sextuplets, but in *quintuplets,* which are 20th notes in actual
value. The metric signature could be, logically and precisely,
15/20. But since it is not a unitary meter, and since it is based
on the grouping of pulses, the 3/4 notation is clear, simple and
practical.

EXAMPLE 101

Rhythmicon #89 - Creston

Two relics of medieval metric signatures are noted *en passant* and
should be abolished without any explanations or regrets. These

two are C for 4/4 meter and ₵ for 2/2. They have been consistently misinterpreted as Common Time—4/4—and Cut-Time or Alla breve; whereas, in medieval terminology C = ◧ ◧ meant Duple Meter (2/4), and a vertical dividing line thru the symbol ₵ meant "twice as fast."

As explaind in Chapter I, a Multimeter is one in which the meter is changed continuously or continually in a section or thruout the entire composition. With respect to metric signatures in multimeters, it has been demonstrated in Creston's *Principles of Rhythm* that constant changes of signature are often unnecessary. This idea was suggested earlier with the partial analysis of Stockhausen's "nr. 2 Klavierstücke I-IV" (Example 84). A full analysis of one such specimen of Multimetric Signatures is quoted from *Principles of Rhythm,* the excerpt of Bartok's String Quartet No. 3.

EXAMPLE 102

2nd Movement (Allegro) of String Quartet No. 3 – Bartok

From cue [3] to the first measure inclusive of [9], there are 189 eighths. This is divisible by 3, making 63 measures of 3/8. Logically enough, the ensuing section from cue [9] to [28] is all in 3/8. Why not, then, write the preceding section entirely in 3/8, as follows?

Idem – Bartok

For further demonstrations the reader is referd to *Principles of Rhythm,* pages 143-146.

The unnecessary complexity of multimetric signatures is also evident in Polymetric Signatures. Two demonstrations—one from a classic work (Mozart) and one from a 20th-century work (Ravel)—are presented. In Mozart's *Don Giovanni* we find the following example of Polymeter (Example 103).

EXAMPLE 103

The 3/4 meter is the common-denominator of all three meters: The 3/8 meter is actually 3/12 since it equals a 1/4 note in duration; there are 9 measures of 3/12 which equal 27/12 or 9/4 or 3 measures of 3/4. The 2/4 part contains 4 1/2 measures which equal 9/4 or 3 measures of 3/4. Hence, the three meters can be notated in 3/4 (Example 104).

EXAMPLE 104

The 20th-century demonstration is Ravel's Trio for piano, violin and violoncello, the second movement (Example 105).

EXAMPLE 105

There are 3 measures of 4/2 which equal 12/2 or 24/4 or 8 measures of 3/4. The common-denominator meter is therefore 3/4; so the upper part in 4/2 can be notated in 3/4 (Example 106).

EXAMPLE 106

It must be understood that the foregoing criticisms of metric signatures are not to censure the composer's rhythmic sense, but only the impractical and irrational notation.*

---

*In *Principles of Rhythm,* on page 151, will be found an example of polymetric—or rather, bimetric—notation, (Example 154a), which seemed unavoidable, that is, incapable of being written in a single metric notation. When it was realized that the 6/4 part should be beamd according to the 4/4 alinement, the solution was found. In the following examples, the first (a) is Wagner's notation, and the second (b), the correct one.

One other point must be clarified: when is a Polymeter actually existent in combined meters? Since meter is "rhythm by length . . . and length alone," the various meters present must be of *different lengths*. In the Mozart example (103 and 104), the 3/8 part was proven to be actually 3/12 or 1/4, and 3 measures of it equal 3/4, which makes it the *same* length as the lowest part in 3/4. But the 2/4 is a *different* length and hence creates a Polymeter or, more specifically, a Bimeter.

An example of true polymeter is found in the concluding section of the second movement of Creston's Symphony No. 2 (Example 107).

EXAMPLE 107

Final movement of Symphony No. 2 – Creston

Of the four meters, two are of similar length—12/16 and 6/8; but the 2/4 and 5/4 are not. Note, too, that the 6/8 is not the traditional one, but a true 6/8 which results from a different subdivision of 3/4. The common-denominator meter is 3/4, and the notation simplifies immeasurably the conductor's task. The four meters of Example 107 were arrived at thru the Rhythmic Structures. The 5/4 is derived from the Fourth Structure (Regular Subdivision Overlapping); the 2/4 is also derived from the Fourth Structure; and the 6/8 and the 12/16 are derived from the First Structure (Regular Subdivision). The actual calculation can be learnd from the explanation of the Structures in chapter I.

The first definition of Meter was given as "the grouping of pulses or units in a measure or *a frame of 2 or more measures.* Thus far only single-measure meters—Monometers—and Monometric Signatures have been dealt with. To clarify the phrase, "a frame of 2 or more measures," we shall examine several Multiple meters, such as Dimeters (2-measure meters), and Trimeters (3-measure meters).

Multiple meters are usually, but not always, formd when the entire measure is the actual pulse. For example, Mendelssohn's *Scherzo* (Example 108) is in 3/8 which, as a Monometer, implies 3 pulses in a measure. But because of the pace— ♩. = c. 72—one measure constitutes one pulse, and it is so conducted. Metrically and musically, however, it is a Dimeter = a 2-measure meter.

EXAMPLE 108

"A Midsummer Night's Dream" - Mendelssohn

The hemiola in 3/4 at a fast pace is also a Dimeter (Example 109), thru the Fourth Structure.

EXAMPLE 109

Valses nobles et sentimentales - Ravel

The Scherzo from Beethoven's Ninth Symphony (Example 110) is a Trimeter—a 3-measure meter—and is so plainly indicated by his own notation, "Ritmo di tre battute" (Rhythm of three beats).

EXAMPLE 110

And further on in the Scherzo, Beethoven indicates "Ritmo di quattro battute" (Rhythm of four beats), or a Tetrameter—a 4-measure meter (Example 111).

EXAMPLE 111

The foregoing examples of Multiple meters are of the type in which the entire measure is the pulse. There can be, however, multiple meters in which the entire measure is *NOT* the pulse. Such meters are formd thru the Fourth Rhythmic Structure—Regular Subdivision Overlapping. The *Scherzino* from Creston's Suite for Flute, Viola and Piano (Example 112) is such a meter, a Pentameter or 5-measure meter resulting from the Fourth Structure.

EXAMPLE 112

Suite, Op. 56 – Creston

The virtue of multiple meters is the avoidance of unduly long measures; a 10/4 meter can be replaced by a pentameter, 5 measures of 2/4; and a 12/4 meter, with the quarter note as the pulse, can be replaced by a tetrameter, 4 measures of 3/4.

ITALIAN TERMS AND EXPRESSIONS

Before proceeding to the Classification of Meters—the next chapter—it is advisable to clarify. Italian expressions in rhythm. For the Italian expressions in dynamics, the reader is referd to Creston's article, *Common Errors in Italian,* in the July, 1971 issue of *Music Journal.*

Prepositions, adjectives, articles, etc., in expressions

a, al, all', alla = at, in, to (depending on the noun or idiom)
a tempo = in time   *fine al Tempo I* = up to Tempo I
alla breve = in duple meter (2/2)
il, lo, la, l' = the   *l'istesso* (*lo stesso*) *tempo* = the same pace (tempo)
con = with   *col, coll', coi, colla* = with the
con anima = with soul   *colla parte* = with the part (solo)
di più = more   *accelerando di più* = accelerating more
ancora = still   *ancora più mosso* = still faster
meno = less   *meno piano* = less soft
più = more   *più animato* = more animated
molto = much, very   *con molto spirito* = with much spirit
   molto allegro = very fast
poco = little, slightly   *poco meno mosso* = a little less moving
   poco ritenuto = slightly held back
**a poco a poco* = little by little, gradually

---

*The first 'a' is invariably omitted by American and British editors, but never by Italian editors because it (the omission) is definitely *incorrect.*

OTHER EXPRESSIONS

Definitions in parentheses are literal translations.

accelerando = (accelerating), becoming faster
allargando = (broadening) becoming slower
a tempo = (in time) original pace; indicated after a *ritardando,
    ritenuto,* etc.
calando = (lowering) becoming softer
calmando = (calming down) musically interpreted as *diminuendo*
    and *rallentando*
come prima = (as at first)
comodo = comfortably, leisurely; musically, at a leisurely pace
doppio movimento = (double movement) twice as fast
più animato = more animated
più lento = slower
più mosso = (more moving) faster
rallentando = slowing down. Abbreviated 'rall.'
ritardando = (retarding) abbreviated, 'ritard.' NOT synonymous
    with *ritenuto* (q.v.).
ritenuto = (retained) held back. Abbreviated, 'rit.' The abbrevia-
    tion should never be used for *ritardando*.
rubato = (robbd) a flexible pace accomplisht by retarding, accel-
    erating and retarding within a phrase
stringendo = (squeezing) pressing forward or accelerating. Ab-
    brev. *string.*
tempo giusto = (correct tempo) restoring the strict pace after a
    rubato
Tempo primo = (first tempo); original pace
trattenuto = held back, same as *ritenuto*

INDICATIONS OF PACE

Most indications of pace in music are literally descriptions of
mood or character, and only figuratively, of pace. *Allegro* literally
means 'happy,' but is interpreted musically as 'fast.' Also it must
be borne in mind that metronome markings, despite the apparent

precision, merely give a general idea of the appropriate pace. The smallest unit employd in a meter must be considerd too. For example, in a piece in 4/4 at ♩ = 120, employing half notes as the smallest unit, the pace would certainly not be considered 'fast.' And in a piece in 2/4 at ♩ = 60, employing 32nd notes thruout would not be felt as 'slow.'

In the following list, the literal translation is in parentheses and the musical interpretation follows it.

Adagio: (At ease), slow. *Molto adagio*: very slow *Adagissimo*: as slow as possible. *Adagietto*: rather slow

Lento: the same as Adagio and its modifications: *molto lento, lentissimo, Piuttosto lento* 'rather slow'

Largo: (wide, broad) the same as Adagio and its modifications: *molto largo, larghissimo, larghetto*

Andante: (going—NOT walking—from *'andare'* 'to go') musically interpreted as 'moving smoothly' NOT 'slow'

Andantino: (diminutive of *'andante'*) not moving as much as *andante*; the opposite is the common but erroneous interpretation

Moderato: moderate, neither fast nor slow

Allegro: (happy) fast *Allegretto*: rather fast *Molto allegro*: very fast   *Allegrissimo*: as fast as possible

Vivace: (vivacious) same as *Allegro* and its modifiications: *Molto vivace, Vivacissimo.*

Presto: same as *Allegro* and its modifications: *Molto presto, Prestissimo*

Indications of character or mood which are usually associated with pace.

Agitato (agitated) *Appassionato* (impassiond) *con brio* (with vivacity) *con fuoco* (with fire) *energico* (energetic) *giocoso* (playful) *grazioso* (graceful) *Grave* (grave, serious) *impetuoso* (impetuous) *Maestoso* (majestic) *Marziale* (martial) *Solenne* (solemn) *sostenuto* (sustaind) *scherzoso* (jesting) *Precipitosamente* (precipitously) *Precipitevolissimevolmente* (longest word in the Italian language meaning at 'headlong speed')

EXERCISES

1. Simplify the notation of the following metric signatures by reducing each pair to one signature (the pattern is repeated in each case).

2. Convert the following Sequential meters to one metric signature by either combining the meters, e.g. 4/8:5/8=9/8, or forming several measures of the smaller meter, e.g. (9/8:6/8:9/8) = $\frac{9+6+9}{8}$ =24/8=4 measures of 6/8.

3. Convert the following combinative metric notations to a single notation.

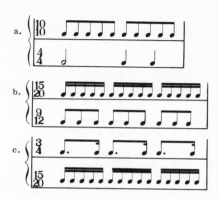

4. Convert the following multimeters to a single metric signature; for example:

$\frac{4}{4}$ $\frac{2}{4}$ $\frac{3}{4}$ $\frac{2}{4}$ $\frac{3}{4}$ $\frac{2}{4}$ $\frac{2}{4}$ $\frac{3}{4}$ $\frac{21}{4}$ = 7 measures of $\frac{3}{4}$

a. $\frac{3}{4}$ $\frac{3}{4}$ $\frac{2}{4}$ $\frac{4}{4}$ $\frac{3}{4}$ $\frac{5}{4}$ $\frac{3}{4}$ $\frac{3}{4}$ $\frac{2}{4}$ $\frac{2}{4}$ $\frac{5}{4}$

b. $\frac{4}{4}$ $\frac{3}{4}$ $\frac{3}{4}$ $\frac{3}{4}$ $\frac{2}{4}$ $\frac{3}{4}$ $\frac{2}{4}$ $\frac{4}{4}$

5. Notate the following combinative meters in one metric signature.

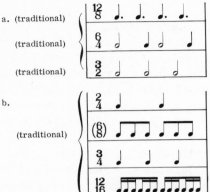

6. Convert the following polymeters to a single metric signature.

VI

CLASSIFICATION OF METERS

Meters can be classfied under three main categories: Monometers, Multiple-Sequential meters, and Polymeters. Each category has several subclassifications. Monometers are 1-measure meters; Multiple-Sequential meters include Dimeters (2-measure meters), Trimeters (3-measure meters), Tetrameters (4-measure meters), and so on. Polymeters include synchronous or simultaneously combined meters, such as 2/4+3/4, 3/4+4/4, and so on.

In the table of Monometers each meter is presented in a paradim showing (1) Duration, (2) Pulses, and (3) Primary units.* Unitary meters, a subclassification, however, show only Duration and Units. The terms 'Duple,' 'Triple,' 'Quadruple,' etc. refer to the *number of pulses* in a measure; the terms 'Binary' and 'Ternary' refer to primary units: 2 in Binary and 3 in Ternary. As an illustration, the meters shown on the first line—2/2, 2/4 and 2/8—are all three termd 'Duple Binary,' as they are virtually the same with only one difference: the value of the pulse as indicated by the denominator. The choice of meter is determind by two factors: the visual-psychological impression and the smallest unit required. For example, 1/2 notes as the pulses give the impression of breadth or slowness, while 1/8 note pulses give the impression of fast pace; hence 2/2 meter seems broader than 2/8. The second factor—the smallest unit required—also determines the choice of meter. If an eighth-note pulse would require 128th notes, it is more practical and convenient to choose either a 1/4-note pulse, making the smallest units 64th notes, or a half-note pulse, making the smallest units 32nd notes (Example 113).

EXAMPLE 113

*Beats are not involved in meter as "the measurement of duration."

93

Most of the monometers diagramd have been in practice since the 16th century, but several, such as 6/2 and 9/2—noted by a question mark—are theoretically possible but impractical. Also, with the metric signatures 6/8, 9/8 and 12/8, a metronome mark is indicated as they are *true* Sextuple-, Nonuple-, and Duo-decuple-*Binary* meters with the 8th note as the *pulse*-value. They are NOT the traditional *ternary* meters.

To clarify the distinction between traditional (6/8) and true 6/8, it will be recalld that the metric signature for binary meter indicates the number of *pulses* in the numerator, and the *value* of the *pulse* in the denominator. For example, 2/4 means 2 pulses, each of 1/4-note-value. In ternary meters, however, the numerator indicates the number of *units* and the denominator, the *value* of the *unit*; in other words, 6/12 means 6 *units,* each of 1/12-note-value. The traditional signature of ternary (6/8) with the 'dotted-quarter' note as the pulse was proven mathematically incorrect in Chapter III, Note-Values. Therefore, if the 1/8 note in 6/8 is the *pulse*, then that 6/8 is a true *binary* meter; this applies as well to true 9/8 and 12/8.

Other instances of true 6/8 meter are: the hemiola in 3/4 (Example 114a), and the metric change from 4/4 to the hemiola in 3/4 (Example 114b). The latter example is misnotated: the First Rhythmic Structure is indicated instead of the Metric Signature of 6/8.

EXAMPLE 114

A word is in order regarding the metronome mark of 60-180 with the binary sextuple, nonuple and duodecuple meters. Metric pulses in music are analogous to the pulse-rate in the human body. If the pulse-rate in a body is well above or below normal, the person is either very sick or dying. Similarly, if the metric pulse-rate in music is above or below normal, the *music* is sick or dying. Maelzel, who constructed the metronome, must have realized this fact when he set the slowest metric pulse-rate at 40 and the fastest at 200. However, in orchestral music requiring 6, 9 or 12, extremely slow conducting beats, it is awkward and difficult to control; hence, the lowest limit is indicated as 60 rather than 40 beats to the minute. For the same reason, the highest limit is indicated as 180—three times the lowest—because conductors, as a rule, combine very fast beats: 4 or 6 very fast beats into 2 or 3 slow beats.

A. MONOMETERS

I. BINARY

DUPLE

TRIPLE

(3/16 and 3/32 are found in 20th-century music. They are unnecessary.)

A unique triple meter which is neither binary, ternary nor combinative, is the basis of Creston's Rhythmicon No. 89, Book 5 (Example 115).* The precise metric signature should be 15/20, as the rhythm of both the melody and the accompanying figuration are based on 20th notes or quintuplets to each 1/4 note. However, the signature of 3/4 is simple and practical.

*This was first mentiond as Example 101.

EXAMPLE 115

Rhythmicon No. 89 - Creston

QUADRUPLE

(4/16 and 4/32 are found in 20th-century music. They are unnecessary.)

SEXTUPLE

NONUPLE

DUODECUPLE

II. TERNARY

Binary Duple, Triple, Quadruple and Sextuple meters have Ternary companions. Ternary Nonuple and Duodecuple, however, are impractical. In the metric signature of ternary meters, the numerator indicates the number of units in the measure, and the denominator, the value of each unit. The signature given is in the Revised Notation, and the one in parentheses is the traditional one.

DUPLE

TRIPLE

*The opening section of Creston's Dance Overture has the metric signature of 18/16 (9/8), because except for one measure in (9/8), the entire section is in traditional (18/16), but is conducted in 3.

QUADRUPLE

Henry Cowell's "Tides of Manaunaun" is misnotated as 4/2; it is, however, obviously in 12/6 (12/4), as the brackets show (Example 116).

EXAMPLE 116

SEXTUPLE

NONUPLE and DUODECUPLE. These Ternary meters are impractical.

It must be noted that ternary meters often have binary metric signatures, and vice versa. For example, 6/12 (6/8) can be notated in 2/4, and 2/4 can be notated in 6/12 (6/8) (Example 117).

EXAMPLE 117

$$\frac{2}{4} \;\; \overset{3}{\text{♩♩♩}} \; \overset{3}{\text{♩♩♩}} \; | = \frac{6}{12}\binom{6}{8} \qquad\qquad \frac{6}{12}\binom{6}{8} \; \text{♩.♩. ♩.♩.} \; | = \frac{2}{4}$$

This noumenon is what makes Combinative meters with a single metric signature practical, and the use of alternating signatures unnecessary. In the table of Combinative meters, the "B" stands for Binary, and the "T," for Ternary units.

III. COMBINATIVE

DUPLE

TRIPLE

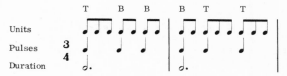

Other possibilities: TTB, TBT, BBT, BTB

QUADRUPLE

Other possibilities: BTTB, TBBT, TTBB, TTTB, BBBT, etc.

SEXTUPLE

Other possibilities: BBBTTT, TBTBTB, TTBBTT, etc.

NONUPLE and DUODECUPLE

These Combinative meters are theoretically possible but impractical. It should be noted that Combinative meters are more conveniently notated with *binary* metric signatures.

In the next classification, the term 'Compound' is used not in the traditional sense but in its true sense. Compound meters are Quintuple and Septuple, such as 5/4 and 7/4; they are generally divided into: 5/4 = 3/4 + 2/4—or the reverse, and 7/4 = 4/4 + 3/4—or the reverse. There are, however, instances in which 5/4 could be termd Quintuple Binary, that is, in which there is no subdivision of 3/4 + 2/4 (Example 118).

EXAMPLE 118

Partita, Op. 12 - Creston

IV. COMPOUND (BINARY)

QUINTUPLE

SEPTUPLE

V. COMPOUND (TERNARY)

QUINTUPLE

The third movement of Creston's String Quartet employs both Quintuple and Septuple Ternary meters: (15/8) and (21/8).

SEPTUPLE

Pulses	$\left(\frac{21}{8}\right)$	$\left(\frac{21}{16}\right)$
Units	21	21
	12	24
Duration		

or the reverse   or the reverse

VI. COMPOUND (COMBINATIVE)

QUINTUPLE

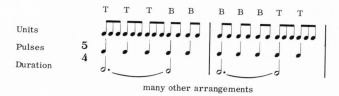

Units	
Pulses	5
	4
Duration	

many other arrangements

The second movement of Creston's Saxophone Concerto has the following Combinative meter

in the orchestral accompaniment, and many different permutations in the saxophone melody.

SEPTUPLE

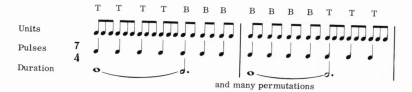

Units	
Pulses	7
	4
Duration	

and many permutations

VII. UNITARY

For all Unitary meters, the metronome setting should be from c.40-c.120 for the entire measure. In other words, the *measure* is the *pulse*.

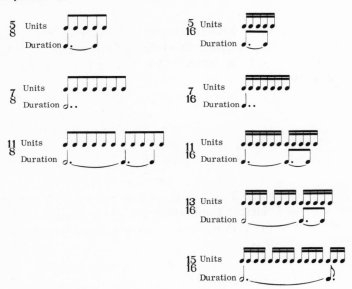

B. MULTIPLE and SEQUENTIAL METERS

In Multiple meters, the entire *measure* is the actual pulse— e.g. 3/4 ♩. =40-120—thereby forming Dimeters, Trimeters, Tetrameters, etc. Such meters cannot be considered Monometers because, as stated earlier, there must be at least 2 *pulses* to a meter; consequently, when the entire measure is the pulse, there must be at least 2 *measures* to a meter.

DIMETER (2-measure meter)

Specimen: 2 measures of 3/8, as in Mendelssohn's Scherzo from "A Midsummer Night's Dream" (Example 119).

EXAMPLE 119

Another type of Multiple meter results from the Fourth Rhythmic Structure—Regular Subdivision Overlapping—in which the measure is NOT the actual pulse. Specimen: 2 measures of 3/4 (Example 120).

EXAMPLE 120

See Examples 24b and 24d in Chapter I for two musical examples of this type of Dimeter. There are numerous examples in music from the 17th to the 20th centuries; also in chapter 6 of Creston's *Principles of Rhythm*.

TRIMETER (3-measure meter)

Specimen: 3 measures of 3/4. The Scherzo from Beethoven's Ninth Symphony contains a section in trimetric 3/4, clearly indicated by Beethoven's own words, "Ritmo di tre battute" (Rhythm of 3 beats).

EXAMPLE 121

Ritmo di tre battute

$\bullet. = 116$

A type of Trimeter derived from the Fourth Structure is 3 measures of 4/4 (Example 122).

EXAMPLE 122

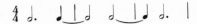

See Example 24a. Chapter I, for a musical example (Palestrina). The following sources of Trimeter will be found in Creston's *Principles of Rhythm*:

> Scarlatti—Sonata, Longo 286, Example 121a, in 4/4
> Brahms—Second Symphony, Example 121c, in 2/2
> Ravel—"Scarbo," Example 121d, in 3/8
> Walton—Viola Concerto, Example 121f, in 2/4

TETRAMETER (4-measure meter)

In the Beethoven Scherzo just quoted is also found a section marked "Ritmo di quattro battute" (Rhythm of four beats).

EXAMPLE 123

It should be noted that in this section dimeters are introduced, twice in the above excerpt and once again 5 measures later.

Tetrameters derived from the Fourth Rhythmic Structure are not uncommon. Following are several sources:

> Mendelssohn—Example 24c, Chapter I.
> From *Principles of Rhythm*:
> Chopin—Valse brilliante, Example 123b, in 3/4
> Ravel—"L'enfant et les sortilèges," Example 123c, in 3/4
> Toch—Piano works, Op. 40 No. 5, Example 123d, in 3/4
> Stravinsky—"Le sacre du printemps," Example 123e, in (9/8)

PENTAMETER (5-measure meter)

Pentameters are more common in 2/4 and 3/4 than in larger meters, 4/4 and 5/4. Following are several musical examples of Pentameters, all derived from the Fourth Rhythmic Structure.

EXAMPLE 124

SEQUENTIAL METERS

Sequential Meters are derived only thru the Fifth Rhythmic Structure (Irregular Subdivision Overlapping). They might be termd "Irregular Multiple meters." Following are several musical examples (quoted in *Principles of Rhythm*):

Sequential Dimeter	Creston—Symphony No. 2, Example 129a, in 3/4
	Creston—Symphony No. 4, Example 129b, in (6/8)
	Ravel—Piano Concerto, Example 129c, in 4/4
Sequential Trimeter	Walton—Symphony No. 1, Example 129d, in 3/4
	Bartok—Concerto for Orchestra, Ex. 129e, in 3/8
	Walton—Symphony No. 1, Example 130f, in 2/4
Sequential Tetrameter	Creston—Piano Concerto, Example 129g, in (6/8)
	Ravel—Trio: piano, violin and cello, Ex. 133a, in 3/4

Sequential Pentameters, Hexameters, etc. are rare, but following are specimens of a Hexameter (6-measure meter) and an Octameter (8-measure meter):

EXAMPLE 125

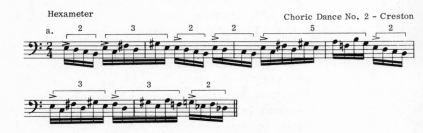

Hexameter   Choric Dance No. 2 - Creston

(Example 125 continued)

C. POLYMETERS

Polymeters are virtually infinite in possibilities. Any kind of complete classification is unnecessary and practically impossible. All that will be attempted here is a definition and several examples.

A Polymeter is one in which two or more different meters are combined simultaneously. The difference must be in length, that is, a meter of two pulses is combined with one of three, a meter of three is combined with one of four, etc. A combination of 2/4 and 6/12, for example, is not a Polymeter but a combination of 2 binary pulses and 2 ternary pulses; while a combination of 3/4 and true 6/8 is of the First Rhythmic Structure and not a Polymeter.

Polymeters need not be indicated by different metric signatures as in the Mozart example—Example 28, Chapter I (q. v.). By means of the Rhythmic Structures, any logical combination of meters can be notated in a single signature. Creston's Second Symphony contains, in the concluding section, four different

meters with a single metric signature, three of which are of varying length (Example 125.) This is the analysis of that polymeter:

2/4 and 6/8	2 pulses
3/4	3 pulses
5/4	5 pulses

The 6/8—true, not traditional—is derived from the First Rhythmic Structure (hemiola); and the 5/4 is derived from the Fourth
Structure (Regular Subdivision Overlapping).

EXAMPLE 126

Symphony No. 2 – Creston

Note that in the third and fourth measures, top staff, the meter is a true 12/16, derived from the First Rhythmic Structure.

Example 30, Chapter I (q.v.), which was presented as a Polyrhythm because of the different configurations of the Third Rhythmic Structure, is also polymetric. Following are examples of Polymeters other than those already noted. They have been selected from Paul Creston's works because they are all in a single Metric Signature.

EXAMPLE 127

Invocation and Dance - Creston

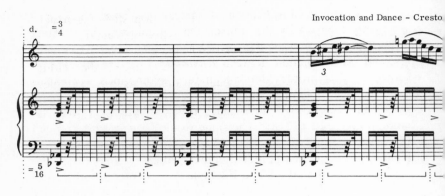

Trombone Fantasy - Creston

The foregoing Classification of Meters may seem to complicate rather than clarify the subject. On the contrary, by insisting on mathematical precision in metric signatures, logical terminology, and the distinction between *Rhythmic Structure* and *Meter,* the choice of a proper Metric Signature is simplified. An examination of the Table of Common Note-Values, Chapter III (q.v.), reveals the notation of note-values in 4/4 juxtaposed with the notation in 12/12 (12/8). The same would be true with 3/4 and 9/12 (9/8), and 2/4 and 6/12 (6/8). Remember that signatures in parentheses are the traditional nomenclature.

The reader should refer to Example 33, Chapter II—the "perfect" meter. The precise but esoteric terminology of that meter is 27/36. It can, however, be more simply notated as traditional (9/8) or Revised 9/12, or as 3/4. As 3/4 it requires more numerals than as 9/12 or 27/36. Since the primary units are 3, making it ternary meter, the logical choice of signature is 9/12 or (9/8).

The distinction between rhythmic structure and metric signature must be carefully considerd. Such signatures as 6/8:3/4 (Example 86), $\frac{3+3+2}{8}$ and $\frac{4+2+3}{8}$ (Example 82) are not metric signatures but rhythmic structures. The first signature is the hemiola (First Rhythmic Structure), and the other two are the Second Rhythmic Structure (Irregular Subdivision).

In the Fourth Symphony of Charles Ives (as published in *New Music Quarterly*), we find the most confused and illogical Polymeter (Example 128). Only the various lengths of the meters are given.

EXAMPLE 128

$\frac{5}{8}$	6 measures	$= \frac{30}{8}$
$\frac{7}{4}$	2 measures	$= \frac{28}{8}$
$\frac{2}{4}$	5 measures	$= \frac{20}{8}$
$\frac{6}{8}$	5 measures	$= \frac{30}{8}$
$\frac{4}{4}$	3 measures	$= \frac{24}{8}$

As can be seen in the conversion to 8ths, only the 6 measures of 5/8 and the 5 measures of 6/8 are of the same length, for that segment of duration. If the 6/8 is the *traditional* meter, we can give it its true mathematical signature of 6/12 and convert all the denominators to 12ths; the result is:

$$
\begin{aligned}
5/8 &= 5/12 \dots\dots\dots & 6 \text{ measures of } & 5/12 = 30/12 \\
7/4 &= 21/12 \dots\dots\dots & 2 \text{ measures of } & 21/12 = 42/12 \\
2/4 &= 6/12 \dots\dots\dots & 5 \text{ measures of } & 6/12 = 30/12 \\
(6/8) &= 6/12 \dots\dots\dots & 5 \text{ measures of } & 6/12 = 30/12 \\
4/4 &= 12/12 \dots\dots\dots & 4 \text{ measures of } & 12/12 = 48/12
\end{aligned}
$$

Now we have *three* meters that are equal in total duration: 5/8, 2/4 and 6/8. But it is still *not balanced*; the 2 measures of 7/4 and the 4 measures of 4/4 are of longer duration than the other three meters. Also, there is proof that the 6/8 is traditional meter in the notation at the bottom: (♩. [of 6/8] = about 50).

VII

SUMMARY

1. Meter and Rhythm are not synonymous. To understand the proper domain of meter, we must have a clear conception of the four elements of Rhythm—Meter, Pace, Accent and Pattern—and the function of each.

2. RHYTHM, in music, is the organization of duration in orderd movement.

3. The five Rhythmic Structures are: I. Regular Subdivision, II. Irregular Subdivision, III. Overlapping, IV. Regular Subdivision Overlapping, and V. Irregular Subdivision Overlapping.

4. METER is the measurement of duration. The elements of Meter are: Pulses, Beats and Units. Pace is a factor in the measurement of duration, but Accent is an element of rhythm, not of Meter.

5. The traditional terms, 'Simple' and 'Compound' as applied to meters are imprecise and illogical. They are replaced by 'Binary' and 'Ternary'. Two other types of meter must be acknowledged: 'Unitary' (5/8, 5/16, 7/8, etc.) and 'Combinative' (Binary + Ternary).

6. The terms 'Duple' and 'Triple', used by some theorists in place of 'Binary' and 'Ternary,' are ambiguous. They should refer to the number of *pulses* in a measure, not *units*. Duple meters are: 2/2, 2/4, 6/12, etc. Triple meters are: 3/4, 3/2, 9/12, etc.

7. 'Time Signature' is *not* synonymous with 'Metric Signature'; it is illogical. 'Bar' and 'Measure' are *not* synonymous; but 'Bar' and 'Barline' *are*.

8. 'Dotted-half note,' 'Dotted-quarter-note,' 'Dotted-eighth-note,' 'Double-dotted-half note,' etc. are graphic descriptions and nondefining. See the Table of Note-Values for their precise durational value.

9. Any note of less durational value than a whole-note is a *fraction* of a *whole note*. The three symbols for note-value, other than note-head, stem, flag and beam—which have been ignord—

113

are: the *dot*, the *tie* and the *numeral*. Consequently, there are several methods of notating a note-value (see Examples 41 and 45). The misconception regarding note-values arises from the practice of allowing one symbol to function in varying ways (Example 46).

10. It is not necessary to invent symbols for such note-values as 1/6, 1/12, and 1/10, or uncommon and rare note-values such as 1/28, 1/96 and 15/16. A numeral clearly indicates any irregular note-value. It is unnecessary to re-educate professional musicians in the revised terminology; one can refer to the *written* aspect of irregular or uncommon note-values (PAGE 41, second paragraph).

11. The distribution of note-values in a measure must be according to pulses and/or units. A single symbol should not indicate a fraction of one pulse plus a fraction of another (Example 54).

12. Theoretically, *any* fraction of a whole note can be used; in actual practice, virtually innumerable fractions have been used, from 1/96 to 63/64 (Example 57).

13. Irrational notation of flags and rests places an unpardonable obstacle in the path of the performer (Example 79).

14. The indication of a rhythmic structure is NOT a proper metric signature (Example 82). If the hemiola is the rhythmic basis, it is unnecessary to indicate it in the metric signature (Examples 85 and 86). If the second structure is used as a metric signature, the beams would be metrically incorrect (Example 82).

15. Metric signatures with a numerator of "1" are illogical. There must be at least 2 pulses to a measure, at least 2 beats to a rhythm, and at least 2 units to a pulse or beat.

16. In Combinative meters employing extrametrical units, the Simple-Practical metric signature is preferable to the Complex-Precise (Examples 98 and 101).

17. It has been demonstrated that constant changes of signature are often unnecessary (Examples 84 and 102); and that different signatures in Polymeters are also often unnecessary (Examples 103 to 106).

18. Meters can be classified under three main categories: Monometers, Multiple-Sequential, and Polymeters. The Monometers are: Binary, Ternary, Combinative, Compound Binary, Compound Ternary Compound Combinative and Unitary. The Multiple-Sequential meters are listed in the next item (Number 19). Polymeters, in which two or more different meters are combined simultaneously, are virtually infinite in possibilities.

19. Multiple meters are usually, but not always, formed when the entire measure is the actual pulse. They comprise: Dimeters (2-measure meters), Trimeters (3-measure meters), Tetrameters (4-measure meters), Pentameters (5-measure meters), Hexameters (6-measure meters), Heptameters (7-measure meters), and Octameters (8-measure meters).

20. By insisting on mathematical precision in note-values, logical terminology, and the distinction between Rhythmic Structure and Meter, the choice of proper Metric Signature is greatly simplified.